Automating Manufacturing Operations:
The Penultimate Approach

Automating Manufacturing Operations:
The Penultimate Approach

By William M. Hawkins

MOMENTUM PRESS

MOMENTUM PRESS, LLC, NEW YORK

Momentum Press®, LLC
222 East 46th Street, New York, N.Y. 10017
www.momentumpress.net

ISBN-13: 978-1-60650-367-6   (hardback, case bound)
ISBN-10: 1-60650-367-7   (hardback, case bound)

ISBN-13: 978-1-60650-369-0   (e-book)
ISBN-10: 1-60650-369-3   (e-book)

DOI: 10.5643/9781606503690

Cover design by Jonathan Pennell
Interior design by Exeter Premedia Services Private Ltd.
Chennai, India

10 9 8 7 6 5 4 3 2 1

Printed in the United States of America

Even though full automation of a task may be technically possible, it may not be desirable if the performance of the joint human-machine system is to be optimized.

Dr. Mica R. Endsley (1996)

Contents

List of Figures

Preface

Automation has gone from being a choice to being a necessity to manufacture today's intricate products at the required speeds. This makes automation increasingly a social issue, one that affects people outside the group that designs, controls, and uses smarter, faster machines. The title reflects the fact that it is going to be difficult and perhaps very dangerous, to reach the lights-out factory. Next to ultimate should be as far as we go. Automation systems should be designed to be operated by people.

This book is about past, current, and future manufacturing automation. It is intended for a general audience but will be most interesting to those who work in the field or are considering doing so. There are no equations and no discussions of control theory, and there are no commercial promotions. People, processes, and technology are at the heart of the book.

I wrote the book because I received great satisfaction from translating process designs into operating processes over fifty years as a user, vendor, and consultant. I'd like to pass that enthusiasm on to someone else. This book can serve as a foundation on which to build specialized knowledge for a field of automation.

There is some history to outline how we got to the current situation. The history will probably remain constant as the book ages, but the rest of it won't. Technology changes so quickly that it is pointless to go into detail about current practice. The practice of automation is covered in such a way that you will have the key concepts to be able to use search engines to discover what is now current. The WBF book series from Momentum Press will also help with details.

There is a source of information that stays current, and that is Wikipedia. Note that I have no financial interest in Wikipedia, but I send them money to keep up the good work. Writing this book would have been much more difficult without them. It is the place to go first to get an outline of what you will need to look into further. Scholars do not like Wikipedia because it is open to editing by anyone. I have not found this to be true in the technical articles. Open the Talk tab on any article to see how controversy is moderated. Caution is still advised, as with anything from the Internet. Everything in this book that uses information from Wikipedia has been cross-checked with other sources.

The book is about all three kinds of processes: discrete, continuous, and batch. It also divides processes into fluid and discrete, where fluid processes use pipes to determine material transfer paths. Unique information is presented in Chapter 8 Programming concerning the problem that automation is becoming opaque to the humans that use it. Finally, the last chapter tackles the issues that will be current when you find this book in a library, if libraries still exist.

This book contains some new ways of looking at the way automation engineers do things, as well as critical looks at the big new ideas being introduced by vendors. Some may use the word cynical, but as George Bernard Shaw said, "The power of accurate observation is often called cynicism, by those who don't have it."

The ten chapters and two appendixes are summarized in the following list.

1. Automation requires people who know how to use available technologies to create automation systems. The Prolog chapter briefly summarizes the evolution of people and technology since the beginning of the Industrial Revolution. Very little automation existed before the power sources needed for factories were developed, and then it just kept growing. The rate of growth increased with electricity and exploded with computers.

2. Automation requires processes that can be automated. The Manufacturing Processes chapter opens with some words on matter and energy (essential to any manufacturing process), goes into the wide variety of manufacturing industries that exist in recent times, and then describes some example processes. The examples are taken from discrete, batch, and continuous processes, but they barely scratch the surface.

3. Processes require control for safety and the stability required to produce acceptable quality. The Process Control chapter summarizes the evolution of control capability, divided into the times before and after computers. Many of the control techniques we use were developed before computers. Computers allowed accurate calculations but introduced sampling to the equations. There are many books on control theory that will provide details.

4. Process control requires human operators to handle the shortcomings of process control systems. The Operators chapter describes the evolution from panel boards to control room displays and the things we learned about operating during the transition. The present role of operators is described, with some words about accidents and words about retraining displaced operators that do not lightly dismiss the issue.

5. Processes and operators require management to give them direction for improvement. The Management chapter discusses hierarchical organizations, managing with computers, management styles, the role of ego, and an alternative architecture for the hierarchy.

6. With processes, control, operators, and management covered, it is time to talk about automation itself. The Automation chapter defines it and divides it into three types. The evolution of automation technologies is described because there are processes at each of the five stages today. The last stage is the autonomous process, which has implications further discussed in Chapter 6, Chapter 10, and Appendix B. There are also words about robots, including Baxter, and the dark side of automation.

7. Modern automation requires digital computers, from microprocessors in field devices to server farms. All of those levels of processors must communicate digitally among themselves, and so the Communication chapter was written. Networks, protocols, and the specifics of process control are discussed, along with the differences between fluid and discrete process communications. Foundation Fieldbus and Wireless issues are described in detail.

8. The computers that have become the heart of automation are useless without programs. The Programming chapter discusses the need for open, not opaque, procedures that operators can easily follow. Aircraft have crashed when the pilots couldn't understand what the automation was doing. A proposal for designing such procedures is made, and an example is given. The chapter ends with a discussion of programs, programmers, and their languages.

9. Scientists are driven by curiosity to know how the world around them works. Technologists look for ways to adapt science to something with economic value. Engineers are driven to find out how to implement technology. The chapter on Engineering discusses why people become engineers, and then has a detailed example of how a process control system was engineered.

10. There is more to say about automation, but books have size limitations. Automation, processes, and companies can't exist without infrastructure but there's no room. There is only room to look at some future possibilities for manufacturing, computing, and automation. They include nanomaterials, additive manufacturing, cloud computing and big data, robots, security, real artificial intelligence, and genetic programming.

A. The detailed list of industries that was briefly discussed in the Processes chapter is expanded in Appendix A.
B. There is much more to say about artificial intelligence than could be presented in Chapter 10, so it is presented in Appendix B. There you will find explicit concern for creating machines smarter than we are.

Teachers make the difference in how a life turns out, especially the persistent one: experience. I'd like to thank all of my teachers for making this book possible—mentors, companions, and adversaries—but the list is too long and I'd forget some that should have been remembered. These three were most influential in the field of automation: Richard J. Lasher, Lynn W. Craig, and Robert B. Britton. If you feel left out, write to me and I'll fix it in the second edition, if I live that long.

Learning is essential, but not enough. It is also necessary to teach. I'd like to thank my students for making life worthwhile but I don't know who learned what—outside of my children, and not even then. This book was motivated by the need to teach that afflicts some old people and was supported by my learning companion of thirty years and loving wife Joellen.

I must thank Bill Wray for access to the Channelview control room to talk to the operators there experiencing modern corporate conditions. This book would not be possible if Dennis Brandl hadn't asked me to edit the WBF books, which led to meeting Joel Stein of Momentum Press and the offer to do a book. Millicent Treloar as editor has kept me going as my enthusiasm for writing drooped. The book really would not have been possible without the production crew at Momentum Press.

Thanks in advance for buying the book, and especially for any comments you wish to send to me at wmh@iaxs.net.

Bill Hawkins
August, 2013

Keywords

artificial intelligence, automation control systems, automation engineers, automation procedures, automation technologies, automation war stories, big data for automation, cloud computing for automation, evolution of manufacturing automation, future of automation, industrial control, intelligent machine, machine control communication, process automation, process examples, process management, process operators, virtual machines

About the Author

William M. Hawkins is retired from Rosemount's Controls Division (now part of Emerson Process Management), a former board member of the World Batch Forum, a founder and principal of HLQ Ltd, and the editor for the Momentum Press WBF book series. Honors include the Emerson Electric Technology Award for RS3 in 1989 and being inducted into the Process Automation Hall of Fame by Control magazine in 2008. He demonstrated the knack for engineering at an early age and received the BSME degree from MIT in 1960. His first job was at the Hercules (Powder Company) blasting cap plant, then a fiber and film plant, and then central engineering managing an instrument laboratory. He moved to Rosemount (Engineering) as the systems engineering manager for the Diogenes control system in 1980, then control architect for the new RS3 system, and then into ISA standards groups for field bus (50), batch (88), and MES (95) as funding for a new control system faltered. He took early retirement in 1999 when the plug was pulled on RS3, started HLQ for consulting, worked for Foundation Fieldbus writing specs and test scripts until 2003, and then did fieldbus consulting for Rosemount until 2009. He edited four books for the WBF (was World Batch Forum) in 2010 just before it was absorbed by MESA.

CHAPTER 1

Prologue

What's past is prologue.

—Shakespeare (The Tempest, 1623)

Those who cannot remember the past are condemned to repeat it.

—Santayana (1905)

Introduction

Evolution is a name for the process of adapting to an environment. This is true of anything that can change, including ideas. Natural selection is a name for the process that allows organisms with genes to adapt to an environment. An environment may change by the movement of continental plates, new weather patterns, population pressure, or human ideas. Continents change relatively slowly; ideas can change quickly.

The human prologue begins with our evolution from ancestor species, in response to changes in our environment. Animals that walk upright appeared about two million years ago, based on the fossil record. Recent genetic data (Science News "Tangled Roots" August 25, 2012) suggest a more complex evolution than had been thought. Neandertals have an uncertain origin, appearing in Europe between three hundred and four hundred thousand years ago. *Homo sapiens* appeared about one hundred thirty thousand years ago in Africa. The exodus from Africa that populated Eurasia and Australia started around fifty thousand years ago and came in hundreds or thousands of successive groups. Modern humans appeared about forty-five thousand years ago as Cro-Magnon in Europe. Mankind didn't get to Amerigo Vespucci's continent (Columbus thought he'd landed in Asia) until fifteen thousand years ago, give or take 20%.

One of the differences between Cro-Magnon and Neandertal was a lighter, more fragile skeleton. Cro-Magnon's skull did not have the heavy, crushing jaw that apes have, so the large jaw muscles shrank making more volume in the brain case. That left more room for a larger forebrain, with structures that allow reasoning. Natural selection favored hunters that reasoned about what animal tracks revealed about the animal, its size, age, condition, and probable destination—and how long ago the animal passed by. This requires more than reason; it requires a way to pass knowledge to the next generation and a memory to retain what was learned.

Recent work on epigenetics (Science News "New Drug Target Companion Prognostic Test for Hormone Therapy Resistance" April 1, 2013, among others) has shown that the parent's environment can alter the genes that are passed to children. Other work has shown that the genes that build brains can change before and shortly after birth (see the works of Carl Zimmer). We are not perfect copies of our parents, nor are identical twins identical. The result is that no two people are alike, although they can be grouped by certain similarities. Note that ionizing radiation is not required to drive evolution.

Agriculture appeared about ten thousand years ago, which ended the life of hunting and gathering that caused people to evolve sharing, cooperation, and fairness behaviors for their survival. About three thousand years after that, we had the first kings and armies required to defend the fixed territory required for growing crops. Storing crops for winter meant that there was food that could be taken without doing anything to earn it. Human predators evolved that grabbed things. Combine that with the drive to compete for power and kingdoms, and here we are—in an environment created by grabbers and opposed by sharers. Some say that the pendulum of history swings from one side to the other, while others refer to the predator–prey cycle.

The Industrial Revolution

The desire for intentional environmental change had a great deal to do with the evolution of technology. Early on, humans had to make tools in order to make the things that could be used for hunting and gathering. Tools were limited by the available materials and the need for hardness that would allow a tool to maintain its shape with use. Rocks were used until about fifty-five hundred years ago, then bronze (copper and tin—the start of metallurgy), and then iron about thirty-two hundred years ago. Wrought iron allowed the creation of durable tools and machines at the end of the Dark Ages.

What follows is a summary of the history of selected segments of what is called the Industrial Revolution, beginning with the use of manufactured engines to replace natural sources of power such as wind, water, animals, and people. Man's inventions were driven in that direction because the driving force was (and still is) commerce. Manufactured things become common when many people want to pay to have them. When something becomes common, the commercial winner is the lowest cost producer. This provides the force that favors the lowest cost sources of energy and production.

New inventions require inventors, who had been rare until the revolution. Inventors are creative people who combine knowledge of materials and their behavior with the ability to see new ways of combining them. It helps to have experience with the materials. Opportunities to gain knowledge and experience were not present for most people until communication improved. When knowledge could be gained from conversations and even books, and projects were started to gain experience, then inventors multiplied. Eventually, many of them became engineers.

Lewis Mumford thought that the Industrial Age was brought on by the invention of clocks in the Middle Ages, and not by the steam engine. He wrote that the clock was the key machine, since it is "a piece of power-machinery whose 'product' is seconds and minutes ...." Well, people needed to know what time it was to get to work in the factory on time.

The First Manufactured Engines: Eighteenth Century

Before manufacturing there was the work of individual artisans who directly bartered or sold their work to individuals. Few accumulated wealth, at first. Manufacturing creates wealth by adding value to natural materials. Personal wealth grows from one person selling things to many people, or, to a lesser extent, working for someone who sells things. National wealth grows from taxation of personal wealth. The evolutionary pressure was there, but manufacturing had to pull itself up by its bootstraps. It takes investors with money to build a manufactury (which became manufactory, which became factory).

The real reason that the Industrial Revolution did not start before 1700 was the lack of a metal strong enough to build machines. Previous iron could be cast, but it was too brittle for use in machines. It was possible to heat a few pounds of iron ore and carbon (and secret ingredients) in a small furnace to the white heat necessary to make high-carbon steel. Swords were made of that steel, called Damascus steel, possibly from India, from about twenty-three hundred years ago.

Skipping over the early growth of inventors, investors, and factories, and the invention of patents, brings us to the rising edge of the Industrial Revolution and the story of the steam engine. The examples selected are a small fraction of the inventions that grew with the technologies of metallurgy and power, once iron improved to where power could be contained in engines.

Building on prior work that did not catch on, a blacksmith named Thomas Newcomen built an "Atmospheric Steam Engine" in 1712 for pumping water out of mines. A low-pressure boiler filled the power cylinder with steam as the pump rod fell back down into the mine. The steam valve was closed and the cylinder was cooled with water so that the steam condensed, leaving a partial vacuum that pulled the piston into the cylinder. A simple pivoted lever raised the pump piston down in the mine, bringing a quantity of water to the surface.

The thermodynamics of the engine were appalling, and it only applied force in one direction, but it was used for the next fifty years to pump out mines.

About 1763, James Watt was asked by the University of Glasgow to repair their model of a Newcomen engine. He discovered the flaws that held the engine to five horsepower (too much cooling of the cylinder) and began work on his own engine with an external condenser. The Watt engine, in turn, was limited because the cylinder was not a tight fit with the piston. A 3/8-inch gap was considered to be a good fit. Watt needed money to continue development, and so he went in search of investors among people he knew.

Iron had been smelted from rocky ore for a long time using charcoal fires. The sulfur in the charcoal made the iron brittle, but it was discovered that hammering on hot iron could make it stronger. Water wheels were coupled to hammering machines and to bellows to reach higher temperatures, but few rivers ran by iron ore deposits. Low sulfur coal was found to be superior to charcoal, but it wasn't near running water, either. The situation cried out for a good steam engine.

Matthew Boulton made his money with "luxury" goods, such as buttons and buckles. He created the Soho Manufactory a few miles outside of Birmingham, a market city in the West Midlands of England. Boulton discussed steam engines for the factory with Ben Franklin in 1771. Watt chanced to meet Boulton at Soho, where they talked about steam engines in detail, about a year before Watt's investor went broke. Boulton became Watt's investor when the London Water Works told Boulton that they were looking for steam engines for pumps.

There remained the problem of the uneven cylinder bore, caused by hand rolling sheets of hammered iron. Enter John Wilkinson, who made cannons by boring iron castings. He needed very low sulfur iron, so he moved from a river to be near a field of low sulfur coal. He had an engine from Boulton & Watt that pumped water, which he moved to the new site to replace the horses that pumped the bellows. Disappointed with the engine's performance, he examined it and found the loose bore problem, and knew how to fix it. The new tight bore greatly improved the efficiency of the engine. Wilkinson became the exclusive supplier of cylinders for Boulton & Watt engines.

Improvements followed, as things evolved, leading to the double acting cylinder and rotary motion, which led to Watt's famous fly-ball governor. That led to the art of feedback control, but more about that later.

Strong iron was necessary for manufacturing iron goods like cannon and engines. Pounding the iron with hammers to make wrought iron was inefficient. Henry Cort ended the hammering by developing a "puddling" furnace that removed carbon without blowing air through the iron, reducing the time for making ton of pig iron from half a day to a bit more than half an hour. A British Lord in 1786 regarded the work of Cort and Watt as more important than the loss of the American colonies.

Watt, Wilkinson, and Cort had the ideas necessary to bring strong iron and powerful (for their day) engines into general use. Their ideas would still be dreams without entrepreneurs and investors to provide the organization and money to make the dreams real. Further, the investors would have found other things to do if Parliament hadn't found a way to provide exclusive rights to inventions for a reasonable amount of time. Those rights, called patents, were purchased by others who had the ideas or money necessary for manufacturing, effectively selling future growth to fund development of the inventions.

Textile Machines

What inventors of that time didn't have was marketing. There were only so many engines needed for pumping water and making iron. The Industrial Age took off when new machines for making textiles created a new demand for iron and engines. The demand for clothing is sort of built into people.

In 1700, all clothing that wasn't made from hides was made at home by people who carded wool into fibers that could be spun into coarse yarn (homespun). True, spinning flax to weave linen was ancient, but this was England. Some of the yarn was dyed, and all of it was woven into cloth on a handloom. The weaving process started a long time ago—looms are shown in ancient Egyptian decorations—but the tools for development weren't available to go any further.

John Kay developed a faster way to run the shuttle back and forth while weaving, and patented it in 1733. People who felt threatened by the change broke into Kay's house in 1755 and managed to destroy one loom. Similar problems faced those who invented agricultural machines like the thresher. Weaving and farming employed most of the working class in those days.

The productivity of weaving improved, but not that of yarn. James Hargreaves invented a yarn spinner with multiple spinners, but the yarn was not strong enough. Still, the local spinners broke into his home and destroyed the machine, forcing him to move to another town. Somewhat later Steven Arkwright, with the aid of a clockmaker (the machinists of the period), built a multiple spindle device that was so difficult to turn that it required the power of a water wheel. It became known as the Water Frame, and it was a natural fit for a good steam engine. Arkwright's machines became the demand that spurred the development of the Boulton & Watt engine for rotary use.

Arkwright then developed a carding machine, making him the major supplier of yarn. His factory became very important to the five thousand people of Derbyshire, said to be the home of the Industrial Revolution, who banded together to repel any attempts to damage the machines. This was a successful adaptation to the changing environment of mechanized weaving.

Increasingly, there were people who felt their way of life was being threatened by the new way of weaving. They fought back in the only way they knew how—by destroying the machines when they could. There was no collective bargaining. The government was more interested in protecting a new source of wealth than in helping the displaced people to adjust. No judgment is implied here—it is simply a description of the environment in which evolution took place.

Not all inventors were able to pass through the gates to wealth. In 1775, Samuel Crompton could not interest investors, and could not afford to patent, a machine that he had worked seven years to perfect. The Spinning Mule improved on the derivatives of a spinning wheel to make a yarn that was silky smooth. Crompton tried to make a living selling the yarn without revealing his machine, but the secret leaked out. Arkwright was able to patent his own version of the mule. At least, no one tried to burn down Crompton's house.

Finally, Edmund Cartwright developed a powered loom and patented it in 1785, but it didn't work well. Cartwright built and modified looms, in the time-honored evolutionary fashion, until 1792. Others took up the work and solved the last problems. Power looms consumed the power of steam engines in ever-growing numbers into the nineteenth century. A person who had never seen a loom until taught to operate one could make three to four times as much cloth as a master weaver. Master weavers had to find other work to challenge them.

Industry Spreads to America

On another continent, America, cotton was replacing wool for fine, soft clothing. The cotton plant produces a boll of fine fiber and seeds. The sticky seeds had to be removed by hand until Eli Whitney invented a form of carding machine that would remove the seeds from the fibers. The machine became known as the "cotton gin," but no alcohol was involved. The word "gin" is Southern for "engine." Cotton trade with England was an important source of revenue for the new nation.

There is one other invention of interest to manufacturing in the eighteenth century. Oliver Evans invented the first continuous process when he mechanized a flour mill with conveyor belts, bucket elevators, and screw conveyor tubes. Grain was dumped into a feed bin and processed without human intervention until it filled a flour bin. The first one was built in Delaware (in the United States) in 1785. Many more were built for flour mills and breweries. Joseph Wickham Roe, author of "English and American Tool Builders" (1916) said of Evans and his mill, "he began a series of improvements which form the basis of the modern art of handling materials."

Technology began to be transferred to America, as skilled people left England for the New World. One such man was Benjamin Henry Latrobe, who came to Norfolk, Virginia, in 1795 at age 32 after his wife died. He later moved to Philadelphia where the city water company was interested in the idea of using steam engines. Latrobe drew up plans for a waterworks of a type common in Europe, with steam engine pumps and buried wooden pipes.

The plan was somewhat daring in that there were only three engines in the country at the time, and they weren't in Philadelphia, nor were there any steam engine foundries. Latrobe had to choose people and teach them how to make Watt engines, patents not being a problem at the time. To those who had heard about steam engine explosions, he said that those problems had been solved and went on to say, "A steam engine is, at present, as tame and innocent as a clock." This was technically correct because it was the boilers that exploded, not the engines.

The waterworks project was late and over budget, but that's government work. Latrobe is more famed for his architectural designs such as the south wing of the US Capitol building. Less well known are his efforts to educate talented men in the arts of architecture and engineering that were available to him. He numbered as close friends Thomas Jefferson, Robert Fulton, Eric Bollmann (chemist), and Joshua Gilpin (papermaker).

From "Technology in America," in the article written by Darwin H. Stapleton:

> Latrobe provides a good example of how new or "hot" technology is transferred. It is curious that the economies of several technically less developed nations have developed rapidly apparently because they adopted the latest technology available from the developed nations, rather than simply adopting the standard, tried-and-true technology. This may be accounted for by taking into consideration the enthusiasms which new technology evokes—especially in the trained technician, but also in people generally—leading to the adoption of innovations even when they are at first costly and unprofitable.

What would inventors do without early adopters?

When the eighteenth century ended, people partied like it was 1799. The world of business and ideas took little notice.

The Age of Engineering

The nineteenth century was the century of steam power and Victorian engineering. It was the Age of Engineering, as new machines and engines were developed and sold to willing customers, most of them businessmen. There is no room in this book to do those engineering marvels justice. The interested reader will do well to find a book on Victorian Engineering, such as the one by L. T. C. Rolt.

The 1800s also saw the development of the machine tools required to build the engines and the machines that were driven by engines. In order to build metal machines, you have to cast strong iron (not possible until Cort's furnace) or forge it into rough shape. Then you need cutting and drilling tools, and tools that can machine a part to its finished dimensions, and do that with enough accuracy that the parts will fit together. Then you need fasteners such as screws, bolts, and nuts with standard threads that allow interchangeability within a size.

There was no shortage of inventors for tools, but each had to wait for the right materials and the tools to build the tool, and the standards to make them fit together. An example is the slide rest on a metal lathe. It holds the cutting tool and slides along precisely straight guide ways. You can't build a precision lathe without a means to make the guide ways.

Wilkinson's borer for cannon and steam engines is probably the first (relatively) precision tool made. Previous boring machines drove an unsupported cutting tool into the metal guided only by the hole it was boring. Roe notes that machine tools went from those around for centuries before 1775 to "substantially what we have today [1916] by 1850," as each inventor improved the previous tool. Roe also notes that what could be done with tools in 1850 could not have been done at any cost in 1820.

Three other things were necessary: experienced machinists to use the tools, standard dimensions to translate drawings to parts, and measuring tools like the micrometer. The size of the king's foot or the length of his arm made dimensions subject to politics. Considerable work went into local and international standards of weights and measures. The British standardized their system of units in the Weights and Measures Act of 1824, discarding some really ancient units. The US Office of Standard Weights and Measures began in 1830 and became the National Bureau of Standards in 1901. The French bureau of standards BIPM was created in 1875 with seventeen nations including the United States. The UK joined in 1884 but retained the inch and the pound, as did the United States. Many manuals of Metric-English conversion factors were published.

Machines that could cut to close tolerances finally made interchangeable parts possible. Munitions makers had long sought such parts because it is expensive to throw away a rifle when a single part fails. Eli Whitney is usually given credit for ten rifles with interchangeable parts in 1790, demonstrated to members of the US Congress. Historians have revealed that Whitney was desperate after losing so much money defending his cotton gin patents, followed by a fire that destroyed his factory. In his desperation to win a contract for ten thousand muskets, he had parts filed to fit just ten rifles. He was completely unable to supply ten thousand interchangeable muskets, but he put the Army off for six years while he used the money to rebuild his cotton gin business. Today, this is done by using cardboard cutouts at a trade show to sell a product that hasn't been fully developed.

Inventions

Since Mother Necessity makes herself known to different inventors in different places at varying times, there is often more than one inventor for a new thing under the sun. History tends to remember the inventor that commercialized the invention—that is, it made something that others wanted to buy.

Where did all of these inventors come from? Natural selection does not work fast enough for them to pop out of nothing in the eighteenth century. One possible answer is that hunters in the hunter/gatherer years developed analysis and creative skills by tracking wild animals and that gatherers developed those skills by learning hundreds of varieties of plants along with their uses in food, medicine, clothing, baskets, decorations, and so forth. Selection added genes for these traits over many years, although they didn't affect everybody in exactly the same way. Further, trackers were trained by experienced trackers and gatherers were trained by experienced gatherers. Genetics provided the ability to understand that training.

What we do is analyze problems with the tools we have and use our experience to create solutions. Then we analyze the results to gain new experience and try again. We couldn't invent steam engines before the eighteenth century (except for Hero of Alexandria) because we didn't have the tools, materials, and demand that developed as things evolved. We build best when we build on foundations. The eighteenth century laid the foundations for mechanization.

Early in the nineteenth century, Joseph Jacquard in France built upon the work of three predecessors to perfect a weaving loom that could do patterns in the cloth. He used punched cards (more likely thin pieces of wood or metal) to pick the warp so that the weft would make one line in the pattern. The cards made the loom programmable, in the sense that the machinery didn't have to be changed to make new patterns. Other looms were faster, but they couldn't do patterns.

Jacquard also met with opposition from traditional weavers. The wooden shoe worn by peasants in France (similar to the Dutch shoe) was called a sabot. Disgruntled weavers would toss a sabot into a Jacquard loom to halt its progress. There are those who say that the act was sabotage. Later, we referred to tossing a monkey wrench into the works.

Steam power for transportation had to wait for high-pressure steam, limited by the strength of boiler, piping, and cylinder materials. The Boulton & Watt engines ran at low pressure and had to be built as stationary engines. Large stationary engines either lifted water (using gravity for the return stroke) or turned large, massive flywheels. The double-acting cylinder was necessary to reduce the size of the engine. Transportation engines had two cylinders acting in quadrature to avoid the flywheel so that one cylinder supplied power as the other came to the end of its stroke.

Development of these precursors led to railroad engines and steamboats in the early nineteenth century. As usual, there are stories of disasters that led to improvements, as is common when engineers have no textbooks. Steamboats made the transition from the paddle wheel to the Archimedes screw propeller. Steamboats soon made the graceful Clipper ships and those who mastered them obsolete, as they carried more goods faster for less money.

Construction of bridges and tall buildings became more cost-effective with the development of reinforced concrete by a French gardener named Joseph Monier. His patent in 1867 was for reinforced garden containers, but soon was applied to concrete structures. The Bessemer convertor of 1854 made iron so strong that it was called steel. The Otis elevator of 1853 rounded out the inventions necessary to build tall buildings.

Transportation by railroad was well established in 1860. Scheduling trains, especially over the long runs in the United States and Canada, was complicated by each town defining its own local noon and deriving local time from that. Sir Sandford Fleming of Canada (who became known as the first Time Lord) designed a set of twenty-four meridians, originating in Greenwich, UK, to define regions of constant time. Crossing a time meridian changed time by one hour. This was adopted by twenty-seven nations in 1884, and the time was called Greenwich Mean Time.

Early Electrical Communications

Communication was further enhanced by Morse's telegraph, first used commercially in 1844. It was used to gain an advantage on stock market trading by knowing a stock price before others found out. High-frequency trading is used to do this today. Bell invented the telephone in 1876, which made multiple telegraph circuits over one wire possible by using different audio frequencies. Eventually, it found use for business communication by voice, and to call home.

An excellent book on early communications, one that adds depth to a mere description of inventions, is "The Victorian Internet" by Tom Standage. The Preface begins with these words:

> In the nineteenth century there were no televisions, airplanes, computers, or spacecraft; nor were there antibiotics, credit cards, microwave ovens, compact disks, or mobile phones. There was, however, an Internet.

This Internet was not free, so it was not as widely used as today's Internet. Businesses that could afford it used it for managing remote offices and getting information from providers such as stock markets. There may have been misinformation, but there was no spam.

What follows is one of the more exciting stories of engineering and invention to come from the nineteenth century. It concerns a particularly difficult part of building the Victorian internet infrastructure.

> Many land-lines for telegraph were installed, but water crossings were delayed until a suitable insulator for underwater cable could

be found. That happened in 1849, and a cable was laid from Dover to Calais. Five years later, most of Europe's waterways were crossed, but not the Atlantic. News took ten days to cross the Atlantic on fast picket ships. There was no shortage of investors for the transatlantic cable, due to the time value of investment information, which was the same thing that gave value to the telegraph and Edison's stock ticker.

Cyrus Field, in New York, USA, agreed to fund the project in 1854. After doing the ground work, The Atlantic Telegraph Company was formed with GBP 350,000 in capital. The next problem was to build 4000 km (2500 miles) of undersea cable. That done, two ships set out from Ireland for Newfoundland, but the cable was too weak; it broke 480 km (300 miles) out and lay on the bottom, 3.2 km (2 miles) down. New cable was made, again in two containers, and in 1854 two ships set out to the mid-Atlantic, where the cables would be spliced and the ships would sail to opposite shores.

This failed when the cable broke four times in 800 km (500 miles), and the ships had to sail back for more cable. The third time, in 1859, they succeeded and began sending cablegrams between two continents. The cable failed after 23 days, due to oxidation, marine life, and insulation failure. The last was due do a disagreement about whether high voltage was best for the cable (it wasn't) or low voltage, using the Wheatstone bridge as a detector.

It took six years to replace the capital that had been lost at sea and to develop a better cable. It was decided to abandon the two-ship method and put all of the cable on one ship, but no ship could be found. The greatest of the Victorian engineers, Isambard Kingdom Brunel, built the Great Eastern, 211 m (695 feet) long and 23 m (75 feet) wide, to carry 4000 passengers, some of them in high style, around the world without refueling. She was launched in 1858, but a great steam explosion on the maiden voyage kept the customers away.

After six crossings to New York at unacceptable losses, she was finally sold to a group of investors, who in 1864 leased her for the cable crossing. The cable expedition set out from Ireland in 1865 with the ship riding very well in the water. Nine miles out, the cable testing apparatus indicated a short. They retrieved the cable and found a cut piece of wire driven through the cable. The cable was repaired, but five days later, it happened again, proving that it was sabotage by someone aboard, perhaps hired by one of the picket ship companies or a competing cable company. Sabotage was never proven. This was a very high stakes game.

The cable was repaired and they got under way, but four days later the cable broke near the ship, in 3.2 km (2 miles) of water. They ran out of mooring ropes after the third one broke in attempts to snag the cable on the bottom, and left markers to find the spot again. They returned to England and a dock large enough for the Great Eastern, docking 15 days later. The cable was improved again, and a new batch made for the complete crossing in addition to the cable left after the break.

The fifth expedition left on Friday, July 13, 1866. The crew was warned that saboteurs would be tossed over the side without trial. The only major threat to the crossing was a big iceberg, but nothing broke, and they delivered a working cable to Newfoundland two weeks later. Then they went back to the broken cable with specially designed equipment, found it once and dropped it, then raised the broken cable a month later. The second transatlantic cable in the world was complete six days after that. Cyrus West announced that the two continents were joined, "for better or for worse." The Great Eastern went on to lay cables elsewhere in the world.

So it goes when entering unknown technology areas. Stories like this, but with lower stakes, were repeated throughout the nineteenth century, and indeed longer than that. The NASA spaceflight program entered a new frontier cautiously, and still lost astronauts to unforeseen problems.

Energetic Inventions

A new source of energy to replace coal (gasoline) was developed in 1850, but wasn't practical until oil from the ground was discovered in Titusville, PA, by Edwin Drake in 1859. At first, most of the oil was refined into kerosene for lamp oil. Several internal combustion engines were developed, leading to use of the engine in an automobile by Karl Benz in 1885. Benz did not invent or discover benzene, which was used in gasoline, but that's another story. Rudolf Diesel demonstrated his engine in 1890. Henry Ford introduced the Model T in 1908.

Another new source of energy, electricity, was developed in the nineteenth century. Thomas Edison was the first to distribute electricity for the purpose of adding hours to the day with electric light. The demonstration took place at his lab in Menlo Park, NJ, on New Year's Eve in 1879. The generators for the first electric utility were at Pearl Street Station in New York City, NY, which initially served fifty-nine customers with 110 volts DC for lighting. But it was Tesla and Westinghouse who developed alternating current and the three-phase motor that successfully challenged steam engines for delivering rotary power at the point of use. Evolution has now gone beyond piston steam engines to great turbines that generate electricity to run the electric motors that replaced factory steam engines with their power distribution systems of overhead shafts and leather belts.

The Second Industrial Revolution

Much of the Industrial Revolution took place in the United States and United Kingdom, with developments in Europe. There was trade with China, but Asia had largely ignored the revolution, until Admiral Perry was sent to Japan in 1853 to obtain a treaty that would allow the US navy's coal-consuming, steam-powered fleet to refuel with Japanese coal. The treaty was signed, and then Britain, France, Russia, and the Dutch intimidated Japan into signing more treaties to their disadvantage. This led to the overthrow of the Tokugawa shogunate in 1867 and the establishment of the Meiji Era (enlightened rule). Japan caught up with the revolution in just forty-five years by making a national effort to learn all there was to know.

The first uniquely Japanese invention was a mechanized weaving loom in 1890 by Sakichi Toyoda, called the father of the Japanese industrial revolution by some historians. In 1902, he invented the concept of Jidoka, westernized as autonomation (autonomous-ation), which means that the machine stops itself when it senses a problem. He first applied this to his loom so that it stopped when a thread in the warp or weft was broken. Toyoda went on to found Toyota Industries, making looms and later cars, and applying Jidoka to the manufacturing machines.

Somewhere after 1850, the Second Industrial Revolution started as the use of steel replaced iron. Chemical, petroleum, and electrical industries appeared, and then the automobile. Leadership shifted from Britain to the United States and Germany. US Steel, Bayer AG, Standard Oil, and Edison General Electric were created before 1900, as were the great railroad companies. Other companies include Western Union, Westinghouse, International Business Machines, and Bell Telephone, which became AT&T.

Germany was fragmented into thirty-nine independent states until they were unified in 1871, which delayed their part in the industrial revolution. They were able to learn from the experience of the British and Americans without making expensive mistakes, so there were resources to invest in research, particularly chemistry and electricity. Investments were also made in railways, coal, and steel. Soon Germany joined Britain and the United States as a major source of new developments.

The market for dyes was still open, and German chemists took advantage of that opening. German dyes from Bayer, Hoechst, BASF, and others dominated the market in 1900. By 1915, they had 90% of the market with 80% of products being shipped to other countries, which did wonders for their balance of trade. The companies expanded into pharmaceuticals, photography, fertilizers, pesticides, and other primarily chemical industries.

As a side note, the dyes weren't just used for pretty colors. Some dyes were able to selectively stain biological cells, making it much easier to see what was going on with a microscope. Great advances were able to be made in medical research.

The nineteenth century ended in an environment of increasing change, which had evolved many new ways of manufacturing. The growing complexity of manufactured goods caused separate manufacturing of interchangeable parts to be assembled elsewhere. Supply chains were born. The number of patents required to assure a business freedom from competition created a bureaucracy, with courts and lawyers to settle disputes. Capital financial markets evolved. Government regulations became necessary for stability. The number of people put out of work by new machines dropped as displaced workers found new jobs. At the same time, the population of England went from a steady 6 million in 1700 to about 8.5 million in 1800 to about 30 million in 1900, mostly because of agricultural and transportation improvements, and the inventions that provided jobs for most of those people.

The Twentieth Century

The twentieth century led off with radio, initially called wireless telegraphy until Bell's carbon microphone and vacuum tube amplifiers made voice broadcasting possible. Advertisers loved the idea of broadcasting, and so advertising money fuelled the development of the combination of programming and ads that we have today, which began in 1922.

The automobile transformed personal transportation, created many improvements in production processes, and started the oil companies on their way. Ford's assembly line became the first continuous processing line for the discrete manufacturing industries. Just-in-time manufacturing appeared in Japan in 1938 but was severely impacted by the destruction of suppliers during WWII. It resurfaced in 1980 as Kanban, the pull system that allows zero inventories, which grew into lean manufacturing.

Nuclear fission was discovered in the early thirties, which led to the atomic bomb and nuclear power plants. There might be many more power plants had the bomb not been used. Nuclear reactors have their own control problems, which won't be discussed in this book. Radioisotopes continue to do good work where controlled nuclear radiation is required. Medical X-rays also defy the notion that all radiation is bad for you.

The stored program digital computer had various inventors, as usual. It moved toward commercial reality during World War II, after small (compared to De Forest's Audion) vacuum tubes became available to replace relays. The British developed Colossus to break ciphers, but decided that it was too classified to be turned loose after the war. Other machines were built that did not exactly become household acronyms—MADAM, EDVAC, ENIAC, EDSAC, MANIAC, and UNIVAC. Well, maybe UNIVAC did become well-known.

An excellent book on the era is "A Few Good Men from Univac" by David E. Lundstrom, MIT Press, 1987.

Computers were too expensive to buy, like the early steam engines, so they were leased. This put quite a strain on companies that had to recover their investment in development over the time of the lease. When an IBM lease was up, the cluster of machines was sent to a junkyard to be broken up so that there was nothing that could be used to build a computer. A few people near the junkyard in Kingston, NY, did manage to build functioning computers. Since an IBM 704 central processor had about 400 plug-in units, each with eight or less tubes drawing 4 watts, the only practical use for home computers was to heat the house.

The Age of Computing Machines

Things changed in 1948 with the transistor, which shrank the size and cost of the tube plug-in unit by a factor of about 20. Power consumption dropped, too, but you still couldn't buy one. Hardware such as printers and disk or tape storage shrank more slowly. In 1980, 10 MB of removable disk storage required a stack of dinner plate size disks and a machine the size of a small desk to read and write them.

Things really changed with the integrated circuit, eventually contained in a 14-lead package that made it look like a dead bug when upside down. The chip within the package contained many transistors, which saved the space of all of those single transistor cases and printed circuit traces. There followed an amazing era of miniaturization that led to the devices we know today, at a rate according to Moore's Law. Program and data storage went from threaded magnetic cores to integrated circuits with equally amazing increases in bits per unit area and decreases in cost.

The era of computers for everyone started in 1970, with Intel's 4004 microcomputer. Previous integrated circuits performed logic that was built into their design and couldn't be changed. The microcomputer, like its huge ancestors, executed programs that were stored in external memory chips. It was only a matter of time before microcomputers outstripped mainframes in computing power and made the cost low enough for the millions of people who wanted them.

The innovation that surprised an industry of proprietary system manufacturers was the open architecture standard for IBM PCs. This allowed anyone to make a part that would operate in an IBM PC. It also started the software driver industry. The things that could be done with cards invented by other people caused PC sales to take off. Apple remained closed and prospered, although not with the popularity of the PC.

So far, computers must be programmed by humans who translate human requirements into something that a computer can do. Complex projects can fail because some of the user's problems weren't understood by the programmers. On the other hand, users who don't understand computers can create impossible requirements. A project requires two-way transfer of knowledge to the complete satisfaction of both sides before the first line of code is written. Then the project manager has to deal with scope creep, which is the addition of features as they become obvious to those who didn't understand the limits of computers.

Virtual Machines

Before there were virtual machines there was virtual memory. The machines of the sixties used magnetic core memory, which was assembled by manually threading fine wires through tiny magnetic toroid cores, which made the assemblies very expensive. Memory was also available by reading and writing tracks on rotating drums. The drums were much slower than the cores, mostly because the drum might have to make a full rotation before it came to the desired segment of the track. Still, a useful and less-expensive computer could be built if programmers wrote the contents of a section of core to the drum to save it until it was needed again, meanwhile using that section of core for something else. As technology progressed, it became possible to implement hardware memory managers to swap between static and rotary storage so that the programmer was not involved in the process.

Computer technology evolved swiftly in the early days, to the point that programs that ran on an earlier machine would not run on the new machine that made the old one obsolete. This was a time of growth in a computer's instruction set, as marketing departments participated in an instruction set arms race. The revolution against complex instruction set computing (CISC) came in the form of reduced instruction set computing (RISC, an unfortunate acronym). The reduction came from breaking up complex instructions that did many things from one instruction into more atomic instructions such as load and store. The PDP-11 minicomputer introduced the move instruction as a combination of load and store. You can see how programs became incompatible with hardware.

Programming languages were developed that allowed people to specify what a computer would do when it executed a program. COBOL (COmmon Business-Oriented Language), FORTRAN (The IBM Mathematical FORmula TRANslating System), and BASIC (Beginner's All-purpose Symbolic Instruction Code) were common in the early days. Programs written in COBOL and FORTRAN had to be processed by a compiler to generate binary code for a specific computer. There was no guarantee that the compiler for a specific machine would handle all of the statements in the program until vendor-independent standards for the languages were developed. BASIC was not compiled. The program was read by an interpreter that converted program statements to machine instructions. The interpreters were also specific to the type of computer. BASIC was closer to a what-you-see-is-what-you-get language than a compiled language, which is important for beginners. The programs were not necessarily simple.

The solution to intractable incompatibility was the concept of a virtual machine that used the capabilities of the new machine to emulate the operating system and environment of the old machine. This necessarily slowed down execution of the old program, but processor speeds were increasing so fast that one hardly noticed. IBM introduced virtual machines in the sixties and developed them into something more than emulators. Today there is a layer called a hypervisor (so much more than a mere supervisor) between the hardware of a host machine and a number of virtual machines (VM) each containing an operating system like Windows or Linux with a number of application programs written for each operating system, but unaware of the host.

The hypervisor essentially time-shares the virtual machines with the common host hardware. When one VM pauses to wait for an I/O operation to finish, another VM gets time on the host. It isn't really that simple because today's operating systems time-share different program threads. As machines got faster, and processor chips got more cores per chip, all of that time sharing became very feasible. If the host is loaded to some point, application programs may cease to be deterministic because their timing becomes uncertain. It takes a special hypervisor that schedules the virtual machines to make them predictable. This gives up some of the flexibility of using virtual machines, but there are still many benefits, such as being able to switch a VM from one host to another if a host gets in trouble.

Virtual machines led to the idea of cloud computing, which works well if you don't care where your application program executes or who else might have access to your data. The main advantage is scalability without requiring physical space, equipment, and maintenance in your own facility. Maintenance of cloud computing is business office quality rather than the high availability required by process control. A service may be taken down at any time with a "sorry for the inconvenience."

Digital Communications

The stored program digital computer became yet another tool that changed the world. There were many problems that were considered to be too computational to be solved in a lifetime by human computers with their Marchant calculators. The computer made it possible to create useful applications without building hardware. It was apparent early in the game that isolated computers could be more useful if they could communicate.

Large computer networks were first required by transportation reservation systems in the sixties. A single large computer system took data from entry terminals at airport counters and reservation centers in various cities. The central system was designed so that it would never lose a transaction from one of the terminals. The public switched telephone network carried all of the data because it was the only means available.

A packet-switched network was being developed at that time by the military for secure communication in a network that could tolerate the loss of computers serving as switching nodes. This was not because of atomic bombs—the failure of computers was far more frequent. The network was called ARPANET. Also at that time, big computers that could only do one job at a time were being adapted to multiple users sharing the same machine.

Universities and large businesses drove the need for more users per machine. ARPANET was extended to allow more than sixteen nodes by what became known as the Internet Protocol. A secure means of passing data was called the Transmission Control Protocol, and the combination became TCP/IP, which is the basis for what we call the Internet. Combine the decreasing cost of computing with an expanding network and add people finding more uses for that sort of thing, and you have the environment that drove possibly more innovation than all that had gone before.

Ways to connect to a machine that was connected to the Internet (universities, businesses, and independent service providers (ISP)) went from telephone network modems to digital subscriber lines (DSL) and cable modems. Fiber optic to the home is coming, if the home computing market persists. The hunger for bandwidth also drives innovation.

Human beings being what they are, progress has not been without pain. In the eighties, university networks evolved USENET, which was a large set of organized user groups with a common set of rules called netiquette. Anybody who could get on the Internet could join a group, read the Frequently Asked Questions (FAQ) list, lurk for a while to get the tone of the group, and then post original ideas. Sci.physics.fusion was an excellent source of information during the cold fusion fiasco in 1989. There was a great story about the immortal power supply for a group of DEC computers. Look for it as "VAXen, My Children, Just Don't Belong in Some Places." For a really good time, look up "Usenet oracle" where random people answer the questions of other random people. Some of them were excellent.

Then came another form of "the tragedy of the commons." Companies like CompuServe, Prodigy, and AOL began taking money from people who wanted to be on the Internet in the eighties. Microsoft introduced Windows 3.0 in 1990, which made it possible for non-programmers to do stuff with a graphical interface. AOL introduced an interface for Windows in 1992 and mailed out a flood of AOL disks to get you started for a fee.

AOL added USENET access to its interface in 1993, with no instructions for the use of USENET. The result was called "eternal September" because it matched the trouble caused by college freshmen entering in that month. The freshmen eventually learned the rules, but not the continuing stream from AOL. Those folks proceeded to shout down real scientists with their uninformed opinions.

You could sort of work around it, but the irritation grew. Many of us came from an era where people who didn't know what they were talking about were too embarrassed to show their ignorance and kept quiet—not this crowd. Shame was a foreign concept. Soon the newsgroups became unreadable. People who did know what they were talking about left the group. It still goes on, although AOL no longer carries USENET.

The first spam appeared in newsgroups in 1994 with the green card lawyers, Canter and Siegel. They offered to charge a foreign national to get a green card for employment, which the government would do for nothing. They were bombarded with replies that attached large binary files. They continued to assert that they were within their rights to defraud foreign nationals. They eventually lost their license to practice law, but spam (derived from Monty Python) took off from there. Today, there are many green card lawyers, as your favorite search engine will verify.

The World Wide Web

The World Wide Web began as a means to share documents around the world using the Internet. Tim Berners-Lee wrote "Weaving the Web" to describe the origins. He built the software prototype during the last quarter of 1990, and then tried to sell it and manage its evolution for the next few years. Evolution became rapid as the Internet added people who tried it and suggested improvements.

The system requires a web server that has documents that can be accessed and a web browser that allows anyone to access the documents on the server. The browser uses the hypertext transfer protocol (http) to get information from any computer that serves http documents and display it on any computer that can run the browser software. This had the same effect as IBM opening the access to its PC hardware.

What seemed like a good idea at the time got wildly out of hand. Anybody can put up a web site (server) that may offer truth, lies, or malicious software that takes control of your computer. This environment has made antivirus software companies rich beyond their wildest dreams (fear is the most powerful emotion, competing only with reproduction). Sadly, the anonymity of the internet makes it very difficult to catch offenders, but it has been done.

Web tracking cookies were introduced in 1995. There are times when a web server needs to know the context of the messages it gets from a browser. For example, when you login to your bank you send no cookie. The bank replies with an authentication cookie and the home page. When you click on something in that page, your browser sends a message to the bank with that same authentication cookie. When you log out, the bank cancels your authentication, or if you forget, the cookie will expire after a set number of seconds.

Cookies became very attractive to advertisers. There is as much money to be made by tracking your buying preferences as there is to have advance knowledge of what the stock market will do. Advertisers are driven by the need to attract eyeballs to their ads, but even with pretty girls this is far less efficient than targeted ads to people who have expressed an interest in something. Inquire about tracking cookies and the Adobe Flash player for much more information.

More recent developments in computing will be covered in relevant chapters.

Remembrances for the Future

Santayana wrote that famous phrase in book one of a five volume set "Life of Reason" when he was 42. It is even more interesting in context:

> Progress, far from consisting in change, depends on retentiveness. When change is absolute there remains no being to improve and no direction is set for possible improvement: and when experience is not retained, as among savages, infancy is perpetual. Those who cannot remember the past are condemned to repeat it.

What we can learn from our industrial past is that events required three things to be present: inventors, investors, and an environment conducive to growth.

The environment needed:

- Sources of energy
- Materials such as iron and steel
- Tools and tool makers to shape the materials
- Standards for weights, measures, tooling, and communication
- Science and mathematics to discover new possibilities or show how inventions could be improved
- Patents that granted a monopoly on the use of the described idea for a limited time
- Education and textbooks to reduce the number of wheels that were invented
- People to buy the things that were made, and jobs for the people so that they could buy things

- Advertisers and communications media to show people what they could buy

- Banks to loan money that people had saved, and an interest rate differential that allowed the banks to exist

- Stock markets to raise capital for new ventures

- Government regulations to restrain those who would risk our pension funds or the nation's future

- The national wealth to get it all started.

All of those things evolved from the initial conditions, and here we are.

References

Berners-Lee, Tim. *Weaving the Web*. New York, HarperCollins, 1999.

Kessler, Andy. *How We Got Here*. New York, HarperCollins, 2005.

Lundstrom, David E. *A Few Good Men from Univac*. Cambridge, MA, MIT Press, 1987.

Mumford, Lewis. *Technics and Civilization*. London, Routledge, 1934.

Poirier, Rene. *Engineering Wonders of the World*. Victor Gollancz Ltd., 1960 [Barnes & Noble 1993].

Purcell, Carroll W. *Technology in America*. Cambridge, MA, MIT Press, 1990.

Roe, Joseph Wickham. *English and American Tool Builders*. Yale University Press, 1916, McGraw-Hill, 1926.

Standage, Tom. *The Victorian Internet*. New York, Walker Publishing Company, 1998.

Industrial Manufacturing Processes

Manufacturing is more than just putting parts together. It's coming up with ideas, testing principles and perfecting the engineering, as well as final assembly.

—James Dyson (1947)

Introduction

The word "process" has far too many meanings to be used without a modifier. The default modifier for this book is "manufacturing." A manufacturing process is the set of activities that will create a product by transforming the required materials and energy into the product. It is important that the product be worth more than what went into it if humans are involved. There are natural processes that don't make useful products.

Manufacturing in the general sense requires three things:

1. Locating and extracting natural materials, such as trees, minerals, and petroleum buried in the Earth.

2. Primary processing to convert the natural materials into materials of general manufacturing use, such as iron ore to steel stock, petroleum products to plastic pellets, and beach sand to glass marbles. Steel plates, sheets, and rods can be processed directly, or ingots can be melted for casting in the rough shape of the future product. Plastic pellets are mixed with dye and additives to be melted in extruders to make molded parts and fibers. Glass marbles are melted in furnaces to produce objects, sheets, and fibers. There are many more examples.

3. Secondary processing to convert the primary products into consumer goods that do not need further processing, although some assembly may be required. There are as many kinds of products as there are kinds of consumers.

Natural Resources

We use natural products because it is cheaper than making them. There's a law of physics that says we can't create energy. We depend on things that we didn't make, so where did they come from?

Materials

Every known material thing is made of atoms, which can be arranged in a periodic table of elements. Atoms are made of three particles: electrons, protons, and neutrons. At least, they act like particles in the instruments we use to measure them. We now know that the three particles are made of an assortment of sub-atomic particles that are mostly useful for keeping physicists busy. The particles were probably formed at the origin of our universe, but we won't know for sure until someone repeats the experiment. The science of 2012 says that only 4–5% of the matter in the universe is made of atoms. The rest is about 25% dark matter and 70% dark energy. For details, see an article on the results from the WMAP satellite.

Elements (atoms) have varying numbers of electrons arranged in what are called shells around a nucleus of protons and neutrons. The number of electrons in a shell determines how elements will combine, as discovered by the science of chemistry. The elements that we find on Earth were formed directly from the origin event, such as hydrogen and helium, or created by fusion (nucleosynthesis) in the first stars. Stellar fusion stops at iron, because the fusion of iron consumes more energy than it produces. Heavier elements were created in the explosions of supernovas, which is why the heavier elements are less common. We don't do much fusion here on Earth because of the difficulty of reaching and containing the extremely high temperatures required.

When the Earth formed from the cloud of stardust around our new sun, the high temperatures caused elements to react by electron bonding (chemistry, not fusion) to make compounds like iron oxide. The compounds that weren't gasses were churned up by volcanic activity, dissolved in liquids, or burned in the oxygen around them. The elements have to be separated from their compounds so that industrial processes can make them into other products. First you have to find the original compounds and extract them from their surroundings, which involves drilling, mining, and other separation processes.

Energy

Energy comes from the same stellar events that create elements. Our principle source of energy is radiation from the thermonuclear processes in our sun. Stardust provided unstable matter for the nuclear reactions that heat Earth's core. Late nineteenth century scientists, notably Lord Kelvin, believed the Earth was about twenty million years old based on cooling time, but this was too short for the observed evolution of species. Later scientists discovered radioactivity and our radioactive core. We live on the shell of a nuclear reactor.

Two kinds of energy are kinetic and potential. Potential energy is stored by mass in a gravity field and in elements as either chemical (just electrons) or nuclear (protons and neutrons) energy. If the sun went out and we had no controlled nuclear reactors, we could burn fuel until it or the oxygen was used up. While we are waiting for the sun to go out, we use energy for manufacturing processes.

Kinetic energy is measured as momentum or work. An object with mass requires work to accelerate it from rest to some speed and produces work as it is slowed down. If a weight is raised to a height, its potential energy is weight times height expressed in consistent units. Work is force times distance. Momentum is one half of the mass times velocity squared.

If a weight is dropped from a height in Earth's surface gravity, it loses potential energy and gains exactly as much momentum as it loses potential energy. This leads to the first law of thermodynamics – energy cannot be created or destroyed. There are now four laws, but only the first and second concern work. The first law eliminates perpetual motion machines of the first type – those that produce work without an energy input.

If the dropped weight hits the ground without bouncing, then the initial potential energy becomes heat because the force of the collision produced no movement. This heat is lost to the surroundings and becomes unavailable to do work, and so we have the second law – energy cannot be completely transformed to work because some will always be lost. What gets lost is called entropy. The second law eliminates perpetual motion machines of the second type – those that have an efficiency of 100%.

Entropy defies a clear explanation because it is defined as Boltzmann's constant times the log of the number of possible states of the atoms in the system being examined. The next thing that comes up is Shannon's laws for information systems, and soon entropy is at least as weird as quantum mechanics. Worse, entropy expresses the amount of disorder in the Universe, which can only increase because the number of possible states is never zero.

Don't get wrapped up in over-thinking entropy as contributing to the heat death of the universe. Any entropy we create on Earth is but a drop in the ocean compared to the entropy generated by the Sun, and that is lost in the entropy created by billions and billions of other stars.

There are different forms of energy that can do work: potential, kinetic, thermal, chemical, electrical, electrochemical (battery), electromagnetic (light and other frequencies), sound, and nuclear, to name a few.

Chemical energy as fuel can be burned to produce thermal energy as steam, which can turn a turbine to produce kinetic energy, which can turn a generator to make electricity, which can charge a battery. The energy in a battery can create light from a flashlight or sound from a radio or heat from a laptop computer. So far, nuclear energy is confined to medical uses, generating heat, and obliterating things. Transmuting elements can be done, as for medical isotopes, but is not practical for the lead to gold thing.

Types of Manufacturing Industries

Manufacturing is a classification of industry. The word industry derives from the French word for skill or the Latin word for diligent, both good qualities in an industrious employee.

In economics, an industry is a component in the circulation of money that can be divided into at least three sectors:

- Primary: Extracts products from nature with minimal changes to the product. A great deal of work may go into extracting the product, but natural processes made it. The products generate wealth for an economy because the natural processes do not charge for their services. An economy that runs out of natural resources or loses control of them is in big trouble.

- Secondary: Uses industrial processes to convert primary products into secondary products of higher value. These are the manufacturing industries. Some say you can't have an economy without manufacturing, especially the manufacture of manufacturing machines, but that depends on your definition of economy. The balance of trade for a country is strongly influenced by manufacturing.

- Tertiary: Provides services to an economy. Services do not create physical products so they need manufactured products to function. Some examples are transportation, marketing, and entertainment. Services do not export to the same extent as manufactured goods.

Other groups believe that they should be classified as industries, but there is general agreement on the first three. For example, the class "information services" does not fit well in the tertiary category.

See Appendix A for a detailed list of manufacturing industries. What follows is condensed from that list.

The United Nations has published the International Standard Industrial Classification of All Economic Activities (ISIC), Revision 4, in 2008. The first five sections of the list are:

A - Agriculture, forestry, and fishing

B - Mining and quarrying

C - Manufacturing

D - Electricity, gas, steam, and air conditioning supply

E - Water supply; sewerage, waste management, and remediation activities

ISIC classifies agriculture, forestry, and fishing as primary industries. All kinds of mining and extraction of oil and gas are also primary industries. Other sections have processes that can be controlled, but only Section C Manufacturing will be discussed in this chapter, even though Sections D and E are interesting for automation.

The following is a condensed selected list of divisions of the manufacturing section with their ISIC classification numbers:

10 - Food

11 - Beverage

13 - Textiles

17 - Paper and paper products

19 - Coke and refined petrochemical products

20 - Chemicals and chemical products

21 - Pharmaceutical products and preparations

22 - Rubber and plastics

25 - Fabricated metal products, except machinery (28)

26 - Computer, electronic, and optical products

27 - Electrical equipment

28 - Machinery and equipment

Types of Manufacturing Processes

Manufacturing processes can be classified as continuous, batch, or discrete. Pride of place for the oldest process goes to discrete for stone tools. Indeed, most of manufacturing still uses discrete processes. Next comes batch processing with early fermenting of beer and wine. Continuous processing is the most recent because it takes a lot of consumers to make it worthwhile to produce something continuously. Processes can also be classified as fluid or discrete, where discrete refers to materials that hold their shape and fluid refers to materials that are shaped by their container.

Discrete processes produce discrete parts, which may be identified individually or packaged in statistically sampled groups called lots. The parts are built from drawings with an attached Bill of Materials (BOM). Drawings for simple parts may have notes to the machinist concerning the part, such as knurling part of a knob or chamfering the end of a shaft. A part or product that is an assembly of other parts has a written procedure for the assembly in addition to the BOM. The manufacturing industry divisions 25–30 in the ISIC list above all use predominantly discrete processes. The rest of the divisions listed in this chapter primarily use batch or continuous processes.

Batch processes contain multipurpose equipment and may have flexible transportation among the equipment. The defining characteristic of a batch process is that the necessary equipment is allocated to the processing of one product until the process is finished. The equipment may then be used for a different product with the proper preparation. If the process moves the product in stages through sets of different equipment, then more than one product may be in progress in the larger set of equipment. In the bucket and paddle era, the bucket had to be emptied before the next batch of product could begin. The product assembly instructions are called a recipe, which contains a quantized list of ingredients, value settings for processing conditions, and a procedure to make the product.

In common use, the word batch has a wider range. A machinist may make a batch of parts, but discrete processes are used. A cook may make a batch of cookies, but the individual cookies are discrete.

Continuous processes contain fixed purpose equipment with fixed transportation paths. The flow of materials through the process is only changed to accommodate changes in the analysis of inputs or outputs or changes in the environment. Nothing runs forever, so the process must be started up and shut down at intervals. Operation time does not differentiate continuous and batch processes. A continuous process has material entering while transformed material is leaving, which is not possible in a batch process. A continuous process may be developed when demand for the product can be met more efficiently than with a batch process. Generation of utilities such as electricity, fuel gas, steam, and drinking water are continuous processes.

All of the manufacturing industries require activities that transport materials. Transportation may be a major part of manufacturing within a factory, but it is not a process because it does not transform materials. Short-range transport includes energy and material transfers within a process. Shipping is long-range transport, between plant sites and the world. Transportation within a process can certainly be automated, so it needs to be included in this book. Transportation is in section H of the ISIC list, outside of manufacturing.

All of the manufacturing industries require packaging except those whose products are too large. Packaging machinery is a class of division 28.

Process Categories

First we need to categorize the processes. We will get to process control in the next chapter.

The process to make a product can be complex, so we need a hierarchy to deal with different levels of detail so that we are not overwhelmed by the vast sea of details. ISA-88 presents a useful hierarchy:

Process

Process Stage

Process Operation

Process Action

The noun "process" names the entire process, sometimes with the name of the person who developed it. The adjective "process" is necessary to distinguish common words from their other meanings. A manufacturing facility may have several processes, perhaps of different types. A plant that makes ammonium nitrate fertilizer may have a Haber process to make ammonia, an Ostwald process to make nitric acid and perhaps a Stengel process to neutralize the acid with ammonia and turn it into dry powder.

A process stage is an independent set of process operations and process actions that causes a major change in the materials being processed. The process itself usually goes from stage to stage, but parallel stages are possible. The process of making an automobile has separate parallel stages for making engines and other parts before the last stage of final assembly. The Ostwald process has two sequential stages—burn ammonia in air to make nitric oxide and then dissolve the nitric oxide in water to make nitric acid.

A process operation is an independent set of process actions that cause a physical or chemical change in the materials as required to complete part of a process stage. Assembling an engine is an operation if the engine is only a part of the final product. The people who manufacture and ship engines will consider final assembly of the engine to be a stage in their process.

A process action is a relatively simple activity that is required to complete an operation, such as drilling a hole in an engine block or adding an ingredient to a batch.

The situation is complicated by the different names that have been used for these concepts in different industries. Chemical engineers call a stage a unit process and an operation a unit operation. Discrete manufacturing engineers call everything an operation.

Processes by Type

In order to condense the following list, subcategories of a type that have no second level are shown enclosed in parentheses on the same line.

Discrete

    Cast (sand, resin, die, centrifugal)

    Mold (injection, blow, vacuum, foam, sinter)

    Form (forge, press, draw, spin)

    Separate

      Machine

        Cutting tool (turn, mill, saw, file, broach, plane, drill, grind)

        Chipless (water jet, electron beam, chemical, laser, oxyacetylene torch)

      Shear

    Join (deep welding)

    Condition

      Heat treat

      Mechanical (cold work, shot peen)

      Chemical

    Assemble

      Bond (weld, braze, solder, glue)

      Mechanical fits (press, shrink, keyed, threaded, fastened)

    Finish

      Paint (anodize, galvanize, plate, vacuum metalize, polish)

Batch

    Mix

    React

Separate (distillation, filtering, gravity)

Transfer (fill, transfer, dump)

Continuous   Because continuous processes are derived from batch processes, if only in the research laboratory, the process list is the same, except for:

Reformer

Catalytic cracker

Continuous casting of hot metal

Extrusion of plastic or metal

A Few Examples of Processes

Manufacturing Process for a Scooter

The following description of a scooter is intended to show the things that are normally a part of any process that produces a finished device. The drawings are mere sketches compared to the detail found in production drawings. Dimensions and parts specifications are absent because those details are beyond the scope of this book, as is the title block.

Do not try to build this scooter at home. The design has not been tested for safety.

The plumber's scooter illustrates a generic small-scale manufacturing process with moving parts. The scooter is made from copper pipe and fittings, wood, and purchased wheels. Drawings of parts and subassemblies are shown in Figure 2.1.

An assembly drawing is shown in Figure 2.2.

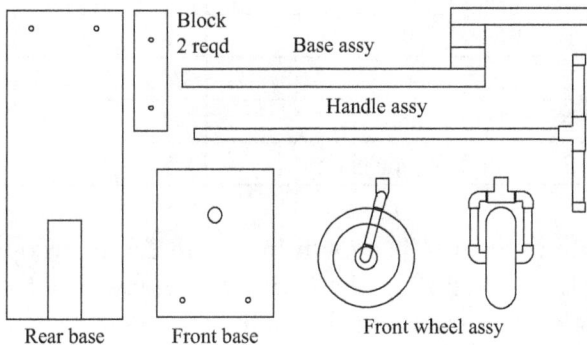

Figure 2.1. Detail drawing.

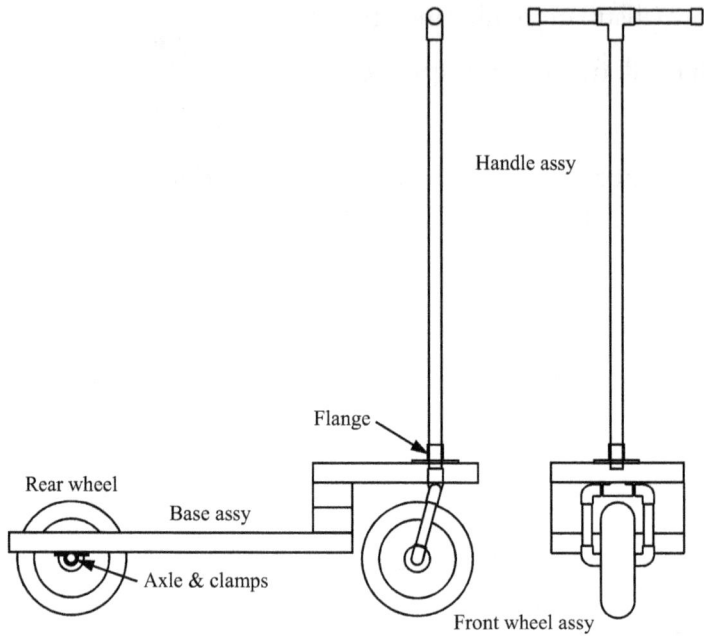

Figure 2.2. Assembly drawing.

The following table is a sketch of a Bill of Materials, with dimensions in inches. Metric standard sizes will be different.

Qty	Description	Size
1	Boxwood plank	$1 \times 4 \times 30$
2	Wheels	5/8 dia. Axle
1	Length of 3/4 copper pipe	60
1	Length of 1/2 copper pipe	16
2	Pipe cap	3/4
4	Elbow	1/2
1	Tee	$1/2 \times 3/4 \times 1/2$
1	Tee	$3/4 \times 3/4 \times 3/4$
1	Deck flange	3/4
2	Surface pipe clamp	1/2

The assembly procedure is shown in Figure 2.3, in an operation process chart. There are three vertical paths listing the operations to be done in the order that a single worker would do them. The final assembly path would normally be below the ends of the three paths, but that would take up too much room in the figure. The numbers in the circles follow one possible order. It is also possible to do all pipe cutting first, then assemble pipe and fittings

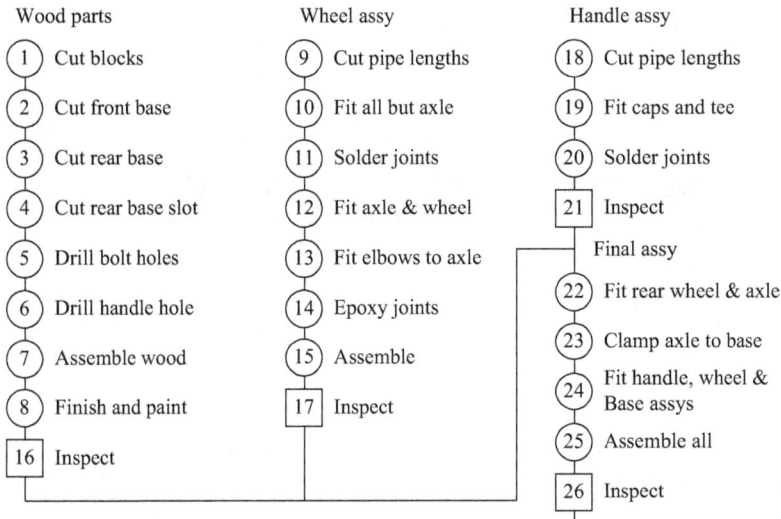

Figure 2.3. Process chart.

with flux, and then sweat solder all fittings. Only the numbering order would change, not the paths. The use of epoxy at step 14 instead of solder is because the wheel is plastic and would not take the heat. The chart is a sketch, not a complete listing of operations. There ought to be a thrust bearing between the front base and the wheel assembly, and the deck flange assembly isn't listed (there is no space for it in the figure).

Manufacturing Process for a Battery Indicator

The battery indicator illustrates a small-scale electronics manufacturing process. The indicator has red, yellow, and green light-emitting diodes (LEDs) to show the condition of a 12 volt lead-acid storage battery. The red LED indicates a bad battery with less than 10.5 volts. Yellow shows that the battery needs to be charged, while green indicates a good battery. Only one LED is on at a time.

The electronic design is shown in schematic form in Figure 2.4. The schematic is transferred to a printed circuit board vendor's schematic capture software. The vendor can then produce a trace layout for the printed circuit with more software, and then fabricate the board and drill holes for parts with a small CNC drilling machine. These days, everything is done with surface mount parts on multilayer boards, with holes only being used to connect traces on different layers. Almost all of the work is done by computer-aided machines.

The completed circuit board with soldered parts becomes a sub-assembly for a box with clip leads that can connect to a battery. The final assembly is packaged according to a packaging drawing that specifies settings for

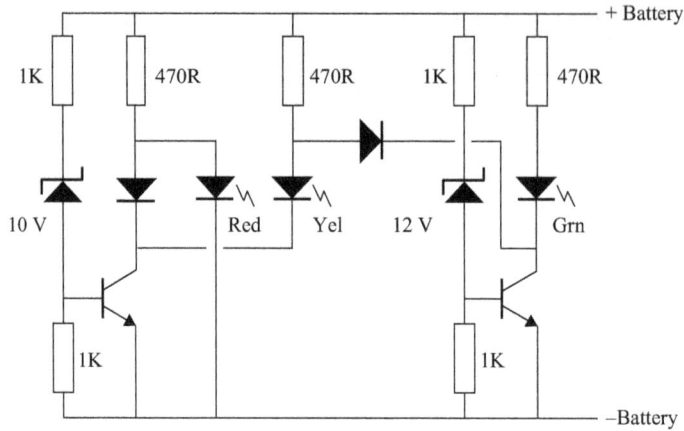

Figure 2.4. Schematic diagram.

a packaging machine and the printing on the label(s). Then it enters the distribution chain on its way to the customer while advertizing creates customers.

All of this could have been shown with the same kinds of drawings, BOM, and chart as the scooter, but here it is left as an exercise for the reader.

Polypropylene Yarn

Polypropylene is very stain-resistant, so the yarn is used for carpets and upholstery fabrics. The color of the yarn is determined by dry mixing polypropylene pellets (from a primary process) with various dye pellets in a cone blender. The mixture then feeds an extruder which forces the melted plastic at high pressure through a screen pack to remove particles and into one or more spinnerets. The spinneret has many fine holes to make fine threads, which are drawn and cooled as they are taken up by a pair of heated Godet rolls. The Godet rolls turn at slightly different speeds to further strengthen the yarn by drawing it at a specific temperature.

The yarn is coated with a finish solution that reduces static electricity build-up and does other things, such as lubricate the yarn through further processing machines. Finish solution also makes a good paint remover and has a strong smell. The yarn is finally wound on a bobbin on a device with two spindles so a new bobbin can be started when the other bobbin is full. A 2" bobbin may be wound out to 12" diameter. Removing bobbins was a manual operation in 1968. Four extruders each fed six spinnerets for 24 take-up stations on a line managed by a pair of operators.

Figure 2.5 shows a line of four spinnerets. Chimneys about 10–15 m high are used for uniform cooling of the strands, which pull together at the bottom. The heated Godet rolls are at a 45 degree angle to save space on the face of the spinning machine. The operator has to catch the falling strands and wrap them

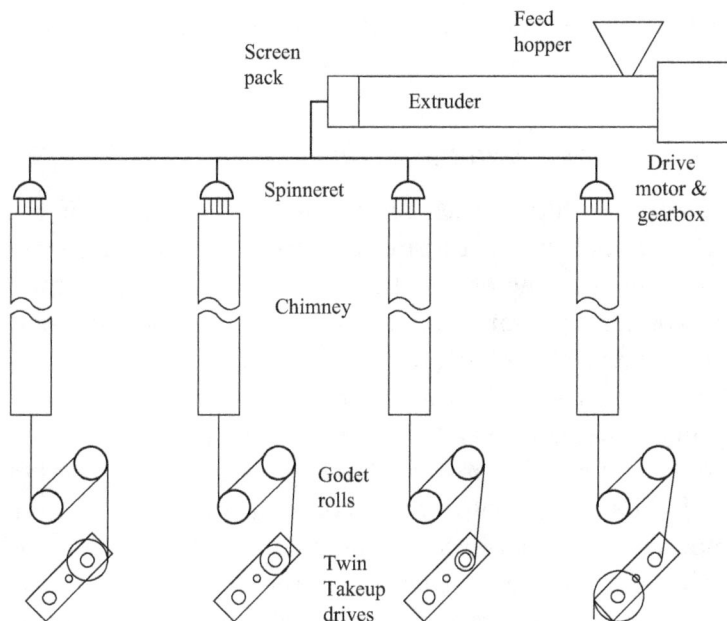

Figure 2.5. Plastic fiber line.

a specified number of times around the rolls to get the desired stretch. The yarn is then quickly wrapped once around an empty bobbin on the take-up device to start a new roll of yarn. The left station in Figure 2.5 has a roll about 75% done, the next 50%, next 25%. At the fourth position, the device has sensed a full roll and rotated one half turn clockwise while cutting the yarn to the full bobbin. The operator will remove the full bobbin and replace it with one that is empty.

An extruder has a hopper to hold blended pellets at the feed (and drive) end of a 3–4 m barrel containing a close-fitting screw. The extruder barrel is heated in 3–6 zones to match the properties of the plastic. Screw speed is relatively low (100 rpm) to avoid excess shearing of the plastic. The outer diameter of the screw matches the inner diameter of the barrel for its entire length. There are three zones along the barrel where the properties of the plastic are changed. The inner diameter is smaller at the feed zone end, where the pellets are heated to their melting point. The inner diameter increases along the melting and compression zone. Then it remains constant through the metering zone where final melting and mixing occur and the pressure builds to about 100 bar against the screen pack and the spinnerets.

Production is continuous from white yarn through the rainbow to black yarn. Then the extruder is stopped while a clean screw is inserted and reheated, and then the process from white to black begins again.

Temperature and draw (elongation) are very important as they determine the orientation of polymer molecules within the fiber from a spinneret hole. If a fan is used to cool the fiber there will be significant differences between the side near the fan and the side away from it. The greatest tensile strength occurs when all molecules are oriented along the fiber.

Vertical Form, Fill, and Seal Packaging Process

Almost all forms of chip snacks, and increasing numbers of other things, are packaged by a machine that forms a plastic bag from a pre-printed roll of plastic. A schematic drawing of a vertical form, fill, and seal machine is shown in Figure 2.6. The machine forms the bag around a hollow feed tube. A sealing device joins the edges of the plastic to form a continuous tube outside of the feed tube, which is pulled down the tube by rollers. A device that seals the plastic across the printed tube and well below the end of the feed tube also cuts the middle of the seal to form the top of the last bag to be filled and the bottom of the next bag. The last bag drops onto a conveyor to the machine that packages bags.

The amount of material to fill a single bag is weighed out and dropped into the feed tube to fill the waiting bag at the end of the feed tube. Time is allowed for the material to settle before the bottom and top sealer do their job and cut the bag free.

Up until 1980 or so, packaging machines were custom designed for each product. A single motor turned a line shaft that coordinated the motions of cams and levers such that one turn of the shaft filled one plastic bag. Bottles and boxes were filled and sealed as a conveyor belt moved them past devices that oriented containers, filled them with a precise amount, sealed them, and applied a label. These were marvels of mechanical design that have to be seen in action to appreciate their functional beauty. Some videos may be found on the Internet.

The computer and the stepping motor made it possible to create motions without cams. Computer programming makes it possible to set up the machine so that different (but similar) packages can be made on the same machine. Cams are difficult to make correctly so that repetitive motion does not break the levers, but all of that knowledge became obsolete when the technology of motion control was developed. So it goes.

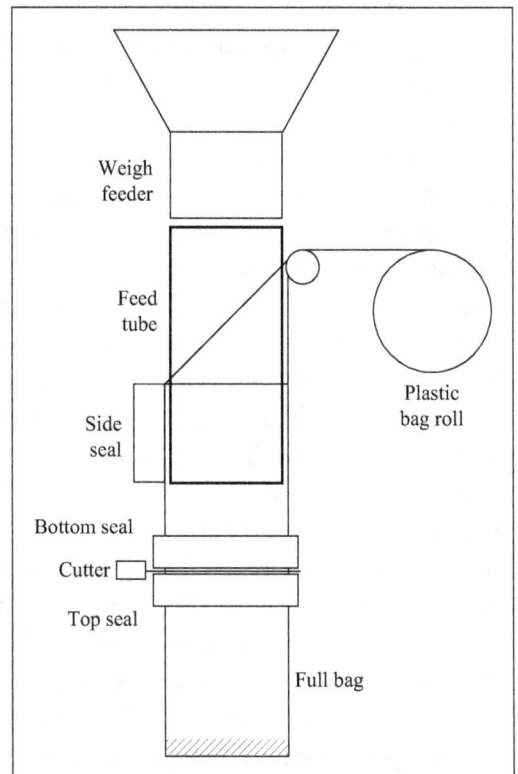

Figure 2.6. Vertical packaging machine.

Wet Corn Milling

The corn starch that is the raw material for many products is processed directly from kernels of field corn. About 56% of the kernel becomes starch, 26% becomes animal feed, and 4% becomes corn oil. The remaining 14% is water. The dry kernels are softened in large corn steeps in a batch process that has a continuous output. The steep water is separated from the corn, which goes to a milling process that releases the germ in each kernel. The germs are separated in liquid cyclones for processing into oil and corn meal.

The remainder is milled and the hulls are separated for drying by screens designed by Dutch State Mines (in 1982). Centrifuging removes the gluten for separate processing and leaves only starch, water, and stuff that needs to be filtered out. The starch may be processed into ethanol, corn syrup (high fructose or not), dextrose sugar, corn starch, chemically modified corn starch, and dextrin. We will only look at the steeping process.

The number and size of steeping tanks are chosen for the desired processing rate. There may be 24 steeps of 500 cubic meters or more, or much less. A steep that has been drained is filled with a solution of water and sulfur dioxide heated to about 52°C and screened corn kernels. Steep water is circulated from the oldest to the newest for the next day or two. The steep water from the newest steep goes to an evaporator because it contains about 6% of the kernel weight. When the time is up for a steep it is drained while the circulation continues. The corn drains into a stream that carries the corn to the first degerminating mills.

Figure 2.7 shows a sketch of four steeps. Draining and filling proceeds from left to right, at an interval determined by the desired steeping time divided by the number of steeps. Steep 3 is prepared by shutting off steep water circulation to it and redirecting that water to a set of evaporators to recover the

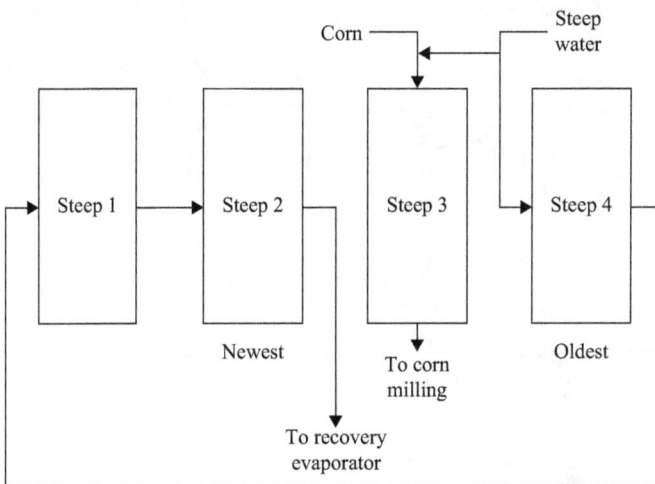

Figure 2.7. Corn steeps.

solids. Circulation flow to steep 4 is also stopped and steep 4 starts fresh steep water. Now the bottom valve of steep 3 is opened, allowing the corn slurry to make its way to the degerminating mills. When the steep is empty, the valve is closed and both hard corn kernels and fresh steep water are added to fill the steep. When the steep is full, it becomes the newest steep while steep 4 remains the oldest. The circulation valves change accordingly. The corn quietly steeps in the circulating water until it is time to drain steep 4.

Polymerization Process

A batch polymerization reactor is like a machine shop, except that it creates parts by stringing molecules together. The raw material is a monomer, a liquid containing unattached molecules, most often hydrocarbons derived from petroleum. More than one monomer may be used in the reaction. After adding some monomer to a tank, an initiator (catalyst) is added to start polymerization, which continues as more monomer is added until the reactor is full. The product may be useful on its own, or may require further processing.

A typical polymerization reactor with a heat transfer jacket is shown in Figure 2.8. Either steam or chilled water flows in the jacket, never both at the same time. The control valve is operated by the reactor temperature controller.

The following is a generic procedure for making a batch of polymer:

Select a jacketed reactor that is ready for use and has the required capacity, heat transfer capability, agitator power, and corrosion resistance.

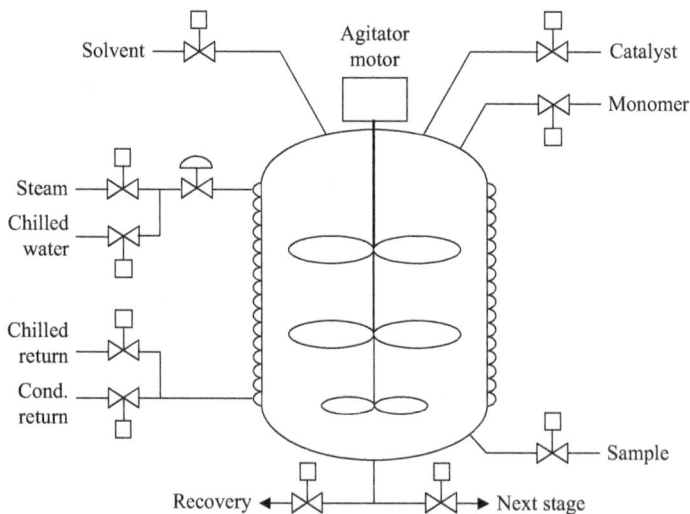

Figure 2.8. Polymerization reactor.

Fill it with enough solvent to cover the first level of agitator blades, start the agitator, and add the initial amount of monomer.

Heat while stirring to the required temperature for initiating the reaction, and then add the initiator.

Wait for the reaction to start as determined by the drop and then rise in temperature.

Add more monomer at a rate that will give the right polymer properties. Cooling will be required to maintain the required temperature. An external heat exchanger may be required.

Add more solvent or reduce monomer flow if the agitator current becomes excessive.

Stop monomer flow if the temperature rise cannot be controlled, as indicated by maximum coolant flow.

Add monomer until the required total amount has been reached, and then stop the flow.

Allow time for the remaining monomer to react.

Cool the reactor and contents down to the required discharge temperature.

Sample the contents for quality, and discharge to recovery if the test fails.

Discharge the contents to the next process stage.

Clean the reactor if necessary, and then indicate that it is ready for use.

Some polymerization reactions are more dangerous than others, depending on the energy released by the reaction. This energy increases with temperature, leading to thermal runaway. The reaction will continue to generate heat until the monomer is consumed, unless the polymer is converting to another exothermic product. This is extremely unlikely as long as agitation maintains a uniform monomer concentration. However, blades have been known to break. Temperatures and pressures may exceed the capability of the reactor to contain them, so a means must be available to prevent explosion, such as an emergency expansion tank.

Some polymerizations are capable of turning solid in the reactor if there is not enough solvent or some contaminant catalyst is introduced. A tank car of liquid polyvinylidene chloride (saran), used to coat polypropylene film for stickiness, will turn solid if an iron nail is dropped into the liquid.

Polymer products include polystyrene, polyvinyl chloride (PVC), synthetic rubber, super glue, epoxies, and nylon. DNA is one of many biological polymers.

Nitric Acid by the Ostwald Process

Nitric acid is made by burning ammonia in air on a platinum gauze catalyst to make nitrous oxide (NO). The gas is cooled and dissolved in water to make the acid. This sounds trivial, but some interesting things are done to recover the heat generated by burning ammonia. The power for the process is mostly from the combustion of ammonia, so the plant must be started with no power from the process.

Air for the process comes from an axial compressor the size of a railroad engine. It has a steam turbine at one end and a gas turbine at the other. At startup, steam from a utility boiler is fed to the steam turbine. At 1% of the rated airflow, a pilot flame is lit and stabilized, and then 1% of the rated ammonia flow is introduced. Combustion spreads a red glow across part of the platinum gauze. When that stabilizes, both fuel and air flows are increased together at a rate that avoids thermal shock to the equipment until the entire gauze is bright at about 900°C. The operator ramps fuel and air up very carefully because a fuel-rich condition could cause an explosion that would blow the very expensive gauze catalyst out of the burner and into the heat exchangers.

As more ammonia is burned, the process heats up. Figure 2.9 shows the heat exchangers and gas paths. The gas flows out to the absorber and back to the expander. A waste heat boiler designed for 900°C nitric oxide drops the gas temperature to something more reasonable for the gas-to-gas exchanger that cools the hot gas for the absorber and reheats the spent gas for the gas turbine, called an expander. The expander produces about 15 times as much power as the steam turbine when the process is at its design rate.

The worst thing that could happen to this process is for the coupling between the expander and the compressor to break, allowing the expander

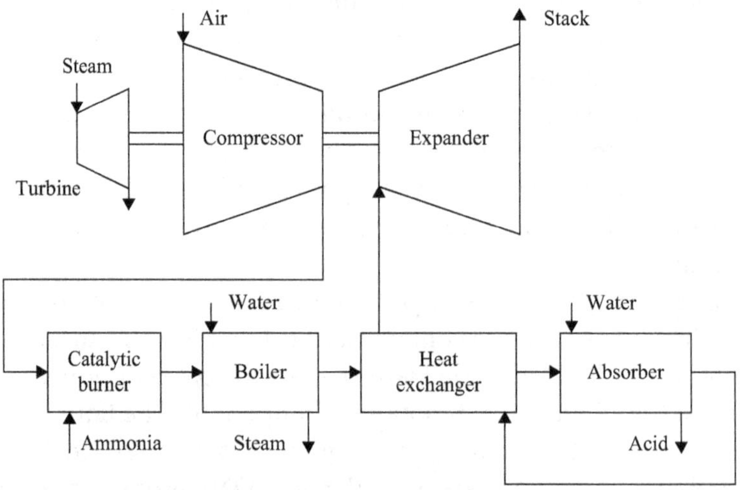

Figure 2.9. Nitric acid process.

to speed up and spray metal all through the plant when it flew apart. A special butterfly valve was built for the expander inlet, powered by bottles of compressed gas and triggered by a simple centrifugal speed switch, with no electricity involved.

The nitric acid is concentrated in a distillation column and then mixed with liquid ammonia in an aluminum neutralizing tank. The tank sounds like a small war was being waged inside it as the chemicals react violently, then the tank drains into a larger tank where acid and ammonia in small quantities are used to neutralize the mixture to the pH of water. The mixture is then pumped to a prilling tower to make ammonium nitrate prills.

The prills could be used for fertilizer, but this plant was in a western Pennsylvania coal region where it was used with fuel oil as a mining explosive. Dynamite had fallen on hard times, but that's another story. The coal from the mine went into a gasification process that produced all of the ammonia used by the nitric acid process.

Titanium by the Kroll Process

Australian beach sand (titanium dioxide called rutile) is burned on coke in the presence of chlorine gas in a fluidized bed reactor to produce titanium tetrachloride ($TiCl_4$ or tickle four), some other chlorides, and carbon dioxide. That's what was being used at a large titanium plant in 1981. Rutile can also be made from ilmenite using a process that removes the iron. Tickle four is a clear liquid between −24 and 136°C, and so it can be distilled to purify it. Not many metals are purified this way.

Titanium is obtained by adding tickle four to molten magnesium at 800°C in a stainless reaction vessel. The result is molten magnesium chloride that is poured off, leaving crystal structures of titanium containing some impurities. The magnesium chloride is sent to electrolysis cells where tens of thousands of amperes keep the mixture molten while chlorine gas comes off and molten magnesium remains. Both products are recycled back into the process. The melting point of titanium is well above stainless steel at 1668°C, so it is mechanically bored out of the stainless container. The metal is then purified in vacuum arc furnaces and cast into primary metal ingots, slabs, and so on.

Figure 2.10 is a schematic diagram of the process. It cannot express the difficulty of handling molten materials and boring out the reaction vessel. The figure shows the reaction vessel in two positions. The reaction proceeds with the vessel upright. When done, it is laid on its side so that the molten magnesium chloride can be drained and transported to one of the electrolysis cells. Then the titanium is bored out of the vessel, which is tilted up to receive another charge. Several vessels are used so that the flow of tickle four is almost continuous.

Production continues as long as rutile, coke, and electricity are available.

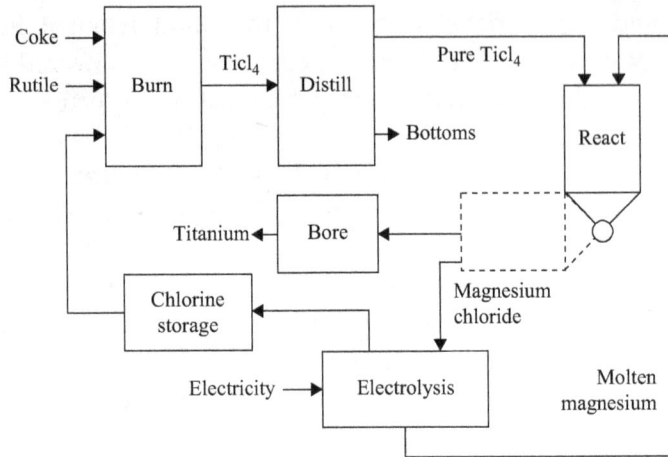

Figure 2.10. Titanium refining.

Note that your version of this process may be different in the titanium extraction details.

Enough

This ends the chapter on manufacturing processes. Many more could have been described, but this book is about automation, not processes.

References

International Standard Industrial Classification of All Economic Activities (ISIC), Rev 4.
Link: http://unstats.un.org/unsd/cr/registry/regcst.asp?Cl=27
Wright, R. Thomas. *Processes of Manufacturing*. The Goodheart-Willcox Company, Inc., 1990.

Process Control

If that guy has any way of making a mistake, he will.
                    —Capt. Edward A. Murphy (1949)

If it can happen, it will happen.
                    —George Nichols, present when Murphy
                    blamed his technician for a failed test.

Introduction

Industrial processes need feedback controls to handle things that happen. A process that is set up and running is subject to disturbances. Some processes will recover by themselves but others need help. The speed of a steam engine is an example, as it could increase until the engine flew apart if the speed was not fed back to the steam supply valve.

Processes involving chemical reactions work best when materials are added and withdrawn according to the conditions of the equation for the reaction. This requires controlled flows. Reaction rates are strongly influenced by temperature, so the transfer of energy must be controlled.

Discrete manufacturing processes are mostly concerned with performing a sequence of operations, which requires control of motion and position. Shaping metals with cutting tools requires control of the cutting speed and feed of the work, and the flow of lubricant.

Control assures consistent quality of the product, which usually increases productivity and reduces costs. These benefits were not immediately apparent. The first engine speed governors were there to keep the machines from destroying themselves. The controls have to be adequate to do the job, and the control system has to be designed for the process and its operators.

Designing for the process implies knowledge of the process, which was slow to come in the days without the process sciences. Control systems had to be designed, tuned, and operated empirically—by learning from mistakes.

When mathematical process models did become available, they were of limited use until the digital computer became capable of using the models on a real-time basis.

The character of the Industrial Revolution changed when electricity became commercially available. Similarly, the Control Revolution (a term that never became popular) changed when the computer became capable of being part of a control system. This chapter will cover some of that, up to supervisory control.

The Control Revolution before Computers

Inventions at the beginning of the Industrial Revolution were not driven by scientific thought. Newcomen would not have built his engine as he did if he had any knowledge of thermodynamics. The inventions were the ideas of practical people who had enough experience to have an idea of what could be done to answer necessity's call to invent something. Many of the inventions were mechanizations of existing processes, building on previous work, and made possible by invented processes for stronger iron.

Scientists of the period were gentlemen who gathered at the Royal Society of London to discuss matters of natural philosophy. Published papers didn't get much further than the Society. It took inventions like steam engines to redirect their thoughts from philosophy to reality. Good science is always preceded by good measurements, and measurements weren't always available in 1700. Maxwell didn't publish his work on the stability of speed governors until 1868—eighty years after Watt's engine had its flyball governor.

And so it was with the Control Revolution. That revolution was driven by the need for closed-loop control in some of the more interesting inventions. The origins were ancient, as were those of textile processes, but first became modern with Watt's governor. What followed was not nearly as well documented as the inventions of the Industrial Revolution, perhaps because it was difficult to see the economic impact of process control. That began to change in the twentieth century.

An Early Control Example

The following story was told by the late Charlie Smoot, a fine gentleman and an expert in boiler control, then at Rosemount Control in 1982. His father had become adept at building machines that could control the steam pressure in a coal fired boiler, but was having trouble finding customers. At the time, about 1915, electric utilities built a number of separate 10,000 lb/hr (4500 kg/hr) coal fired boilers to provide steam for the engines that turned electrical generators. Perhaps the relatively small size was to limit the consequences of a boiler explosion, which was not uncommon in that period.

The steam engines had speed governors that reduced the steam flow when electric load dropped and increased the flow when the load increased. It was necessary to adjust the flue damper on the coal furnace to regulate the steam pressure, especially when the load dropped. You can't reduce the fuel flow with a bed of hot coals, so you have to change the air flow. Smoot had an instrument that would do that, but it was totally proprietary—no standard signals and no chance of using a better damper actuator from some other vendor.

Smoot engaged in negotiations with a vice president of the local electric utility. He promised a demonstration on Armistice Day, 1919, a year after World War I ended. President Wilson had signed a law to celebrate Armistice Day by stopping all business and manufacturing activity for two minutes commencing at the eleventh hour of the eleventh day of the eleventh month (they had style in those days).

Smoot obtained permission to install his system on a set of the utility's boilers. At the appointed hour, all of the electrical manufacturing machinery was shut off. The engine governors closed, the steam pressure rose, the Smoot controllers closed the dampers, and no safety valves lifted, unlike the other boilers. At the end of the moment of silence, the engine governors opened, the steam pressure dropped, and the Smoot controllers opened the dampers to restore electric power. The VP was ecstatic, and Smoot Controls was born.

That might not sound like much, but there was no mathematical method for tuning a feedback controller in those days. Using a screwdriver to change tuning constants was in the future.

Feedback Control

Any process control system requires:

- a sensor to measure a property of the process to be controlled,

- a control device that determines the error between the measurement and a desired target value, and

- an actuator that the control device can move to affect the process to change the measurement.

Drawings of a single control system show the three elements of control and the process arranged in a loop, as in Figure 3.1.

A controller requires an amplifier to increase the error signal to something that can move the actuator. Initially the only amplifier available

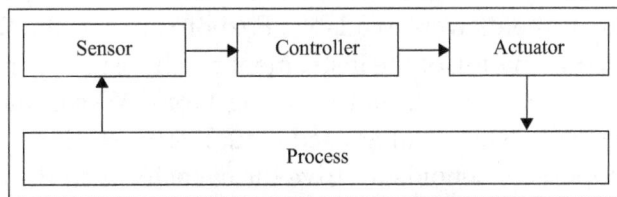

Figure 3.1. Control loop.

to controllers was the lever, as in Watt's fly-ball governor. At the time many pneumatic devices had been built, including the player piano, which included pneumatic amplifiers. When a hole in the piano roll was uncovered, air was allowed to fill a bellows that was otherwise supplied by a partial vacuum, such that a hammer struck the right string and was withdrawn again, without using eighty-eight tiny springs.

A flapper and nozzle device was developed that could turn a small movement of the flapper into a large change in air pressure, as shown in Figure 3.2. A flexible bellows provides the power amplification for the pressure change at a nozzle that is close to a flapper. A restriction in the air supply line provides constant air flow to the nozzle. The pressure at the nozzle is non-linearly proportional to its distance from the flapper. The amplifier was used with a force balance scheme so that the flapper served as a null detector, making the bellows movement very small.

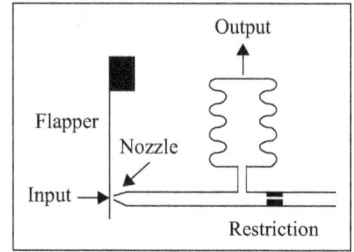

Figure 3.2. Pneumatic amplifier.

The parts of a control loop that are outside of the process are called instruments because they are intricate and precise, as are the tools of medicine and music. The businesses that make process control equipment are called process instrument companies and what they make is called instrumentation. Pneumatic amplifiers are especially intricate. There is nothing that appears to be delicate or intricate in a control valve for pipe one meter (36 inches) in diameter, but the machining of the parts that close together to stop flow is indeed precise.

Vacuum tube amplifiers became available for commercial use in the twenties, but they were not rugged or reliable enough for industrial control use. Magnetic amplifiers were discovered in the early days of alternating current transformers, but they needed further work to compete with pneumatic amplifiers. High gain magnetic amplifiers needed special magnetic alloys that were not available in North America. Patents were filed, but no practical devices were produced for small signal use.

Pneumatic amplifiers were fine for transducers that turned a physical property into a force which could be turned into a pressure. They did not work with thermocouples or platinum resistance temperature detectors, which limited the range and accuracy of industrial temperature sensors. Vacuum tube instruments were available. Foxboro created the Dynalog in the late thirties as a combination of the instrument machinist's art and vacuum tubes. It was used in secret by the military during World War II and released for industrial use in 1946. The Dynalog used a dual variable air capacitor as a null detector and opposed solenoids to drive the capacitor and the attached chart pen.

Germany did not have a shortage of magnetic alloys for magnetic amplifiers. The sensitivity and physical size were improved during World War II as part of the arms race. This work became generally available after the war and was soon used for electronic control instruments. The amplifiers could be small enough to put in a standard instrument case or large enough to control megawatts for drive motors. Power gains of 200,000 were possible. The life of magnetic amplifiers for control designs was cut short by the invention of the transistor less than a decade later.

The output of the error amplifier has to change something in the process in order to close the control loop. Pneumatic amplifiers could cause large changes in air pressure to balance the forces on the summing beams in the controller. Since the output of the amplifier was air pressure, actuators became spring-loaded diaphragms that had the necessary amount of motion and could do enough work to change the position of a fluid valve or a gas damper.

The electrical output of a magnetic amplifier as used in process controllers was too low to operate a valve directly. Electrifying a valve had several drawbacks and made the valve more complicated. The low power wires from the controller output were run to a pneumatic converter box near the valve. The converter used a pneumatic amplifier to balance the pull of a low-power electromagnet against the pressure of air supplied to the valve.

Intrinsic Safety

Chemical processes, especially petroleum refining and natural gas processing, can generate explosive mixtures of flammable gasses. When a gas cloud within the explosive limits of gas to air concentration meets an ignition source, the result can be an expensive disaster. Pneumatic instruments pose no risk—electrical instruments are another matter.

The magnetic amplifiers used by Foxboro were designed for 10–50 milliamp direct current transmission, with current loop power supplies of 65–90 volts DC. This raised concerns for electrical safety in hazardous atmospheres. A great deal of empirical testing based on theory came up with a set of four classes of atmosphere each with limits on the permissible energy stored (in inductance and capacitance) of components in a current loop. This was so restrictive that it was modified by levels of risk based on probability of exposure of the spark to a flammable atmosphere.

The Foxboro energy levels were acceptable for the heavier hydrocarbons, but the transistor made it possible to drop the power to 4–20 milliamps from 24 volt power supplies, which provides a greater margin of safety. Efforts to standardize on 4–20 grew as the number of control components using transistors grew. Foxboro resisted because they had a large user base that they couldn't afford to convert, as well as an investment in magnetic amplifier technology and parts.

The ISA (born in 1945 as the Instrument Society of America) convened a standards committee in 1950 to settle the issue, but it had to be settled by consensus, which wasn't possible in the existing environment. It wasn't until

1973 that 4–20 became the only standard for control signal transmission, and that standard was revised in 1982. Discrete transistors were soon replaced by operational amplifier chips whose performance improved dramatically with time.

Controller Tuning

In a feedback control loop, an error has to exist before the process can be changed. The amount of proportional change has to be just right, as too much will cause the loop to oscillate and too little will yield sloppy control. A delay between the time the actuator moves and the measurement senses that change will cause an integrating controller to oscillate as the integral action continues to work on the past error. The use of too much derivative action will amplify the noise in the measurement, which will wear out the actuator.

Transport delay (dead time) in the process can make it impossible to control at any setting. The following true story should illustrate the point:

> At a paper mill in 1981, an experienced control engineer (but not in boiler control) sat at the computer backup control panel for a 400K lb/hr (180K kg/hr) bark boiler with cogeneration. All of the backup controllers were tuned satisfactorily except the desuperheater temperature for the steam line to the turbogenerator. He put the controller in manual mode and bumped the output up (adding more cooling water spray), so that the step response would give him the dead time. Nothing happened, so he bumped it some more.

> After several minutes of this, the boiler operator asked what was going on because the turbine operator reported water pinging on the blades. The engineer immediately returned the output to its original position as the desuperheater temperature began to drop precipitously. Fortunately, the engineer did not cause any damage to the turbine, and continued his employment with valuable new experience.

A controller had to be adjusted properly, but for years there was no mathematical foundation for the adjustments. Theory might not have done much good with the non-linearities and sticking friction in all moving parts of the loop. These effects would be reduced as machine tools and lubricants got better. It is a tribute to empirical engineering that control loops worked at all in the early years of process control.

Three Control Modes

Feedback control converged on the Proportional, Integral, and Derivative (PID) controller, with air pressure as the signaling system. The first major use of feedback control (not counting engine speed) was for the powered steering mechanisms of large ships. These were servomechanisms that allowed the steersman to position a rudder that a team of ten horses couldn't have moved. Think of the Great Eastern or the Titanic.

Elmer Sperry (of gyrocompass fame) used a PID controller for a ship's automatic pilot in 1911. History gives credit for the first published mathematical analysis of PID control to Nicolas Minorsky, who was working for the US Navy to develop automatic steering systems. He developed the theory from studying a steersman at work as he corrected for present error, factored in past error, and anticipated future error. The work was published in 1922.

Process value sensors became separated from the control devices as new sensors were developed. The valve and its actuator were always in a remote place that wouldn't disturb the sensor. The industry (there was no standards-making body) standardized on 3–15 PSI as the pneumatic signal between the sensor and the controller and between the controller and the valve. A standard signal allowed instrument makers to specialize in sensors, controllers, or valves.

As the twentieth century progressed inevitably toward the future, more control vendors appeared with ever more clever devices to sense process conditions, compensate them, control them to a setpoint, and manipulate the process to change the measurement. At first, necessity drove invention but now competition and the need to get around other's patents picked up the pace. It seemed as though a new instrument and control company appeared with each new invention.

There are too many instrument companies to list here without leaving many of them out. Authors are advised to write about what they know, so the following will be about The Foxboro Company. This company led the field in pre-microprocessor instrument companies with a superior capability in pneumatic control devices. Microprocessors changed the game by making most of the mechanical intricacy obsolete. Instrument makers no longer had control of their process technology.

The company that became Foxboro started making pressure gages in 1909. Beautiful (as clockwork is beautiful) small pneumatic machines of brass and steel were made to do not only control functions but also to do addition, subtraction, multiplication, division, and square root. These only became more intricate as time went on. The ability to build these devices was what differentiated one company from another. Foxboro was very good at it.

Great progress was made, but in 1934 A. Ivanoff of the George Kent Company in England stated that the "science of automatic regulation of temperature is at present in the anomalous position of having erected a vast practical edifice on negligible theoretical foundations." It wasn't until 1935 that the time domain equations for PID control were published by S.D. Mitereff, who named the modes as P, I, PI, PD, and PID.

Tuning the constants of a PID controller empirically was generally not satisfactory. Albert Callander at Imperial Chemical Industries (UK) worked out charts for tuning PD controllers in the presence of significant dead time in 1934, but published internally. The expanded work was published by the Royal Society of London in 1937. The well-known Ziegler–Nichols tuning method for quarter amplitude decay was published in 1942.

Remote Control

Another form of process control was being used that required no tuning constants. Electricity is distributed at various voltages over very wide areas. The big transformers that change distribution voltages are located in substations. High voltage from a long-distance transmission line is reduced by one or more transformers. Each transformer may feed a number of lower voltage (but still high compared to household voltage) transmission lines that go to substations that distribute power to even more substations and eventually to the line that runs along your street.

Each substation may have more than one source of power, but only one source is used at a time. This requires high-voltage switching on both sides of the transformer. There are also instruments that measure the properties of electric power. Switches may be used to add or remove capacitors across the lines to keep the power factor near unity. Some transformers have tap switches to keep the delivered voltage near nominal as loads change.

Figure 3.3 shows a diagram of a substation. Two high-voltage feeders are connected to the two switches on the left side. The other sides of the two switches go to a bus that is connected to the primary of the distribution transformer. The transformer has a remotely controlled tap changer. The secondary of the transformer is connected to a bus that has a capacitor switching bank and three switches for medium voltage distribution feeders. A multiplexor selects signals from the five switches, tap changer, and capacitor bank to send back to a dispatcher. It also sets up commands from the dispatcher to one of the seven devices.

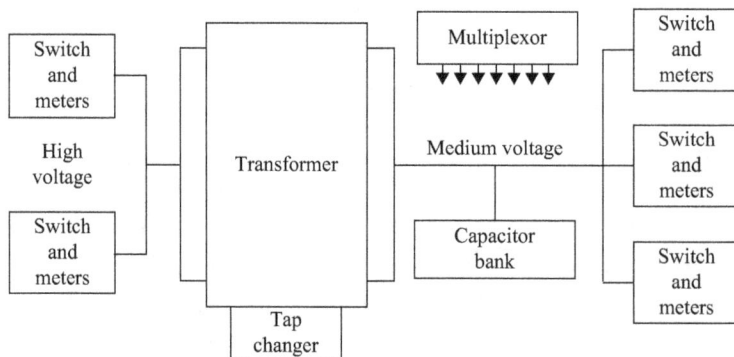

Figure 3.3. Electrical substation.

All of the switching and instrument reading was done manually before World War II. The use of remote control grew during the 1950s. The control was called "supervisory" because it was remote. The term SCADA for supervisory control and data acquisition wasn't used until computers got involved in 1965.

At first dedicated pairs of wires were used for each switch and instrument, but this became impractical. Various means for multiplexing the use of a pair of wires were developed, but none of them assured operation without the chance for an error. The control center sent a command to select a switch at a substation over the pair of wires. The substation replied with the selection that it had made. If the reply was correct, the command to operate the selected switch was sent.

The system of communicating remote measurements and commands is called telemetry. Most people associate that with wireless, but it includes any means such as the telephone network or carrier-current signals over the power distribution lines. Commonwealth Edison in Illinois used telemetry to read instruments in substations as early as 1912.

SCADA is not limited to electric utilities. Any wide-area control system can profit from its use. Municipal water systems, petroleum and natural gas pipelines, even trucks, trains, ships, and planes use some form of SCADA to control the flow of whatever is flowing.

The discussion of SCADA will resume after computers are introduced.

The Computer Revolution

Common use for the word "computer" before 1940 was as a job title for a clerk that operated a Marchant calculator, a true mechanical marvel with things Babbage would recognize and many that he wouldn't. World War II caused the development (an unnatural form of selection) and evolution of the IBM tabulating machines into digital computers with the great survival advantage of stored programs rather than hard wiring. Gone were the days of tinkering with mechanical things. Now it was scientists that designed the first digital computers, using Boolean mathematics and available switching technology.

Scientists were thinking about how to automate calculations in the thirties. Dr. Alan Turing wrote a paper in 1936, which, among others, influenced Prof. John von Neumann to write a report on the design of electronic discrete variable automatic computers in 1945. The key idea was to store programs and data in the same addressable memory. The result was an architecture like that shown in Figure 3.4. Notice the direct memory access connections for the peripheral devices. The Control Unit had connections to all of the other functions to control the flow of data in and out of memory.

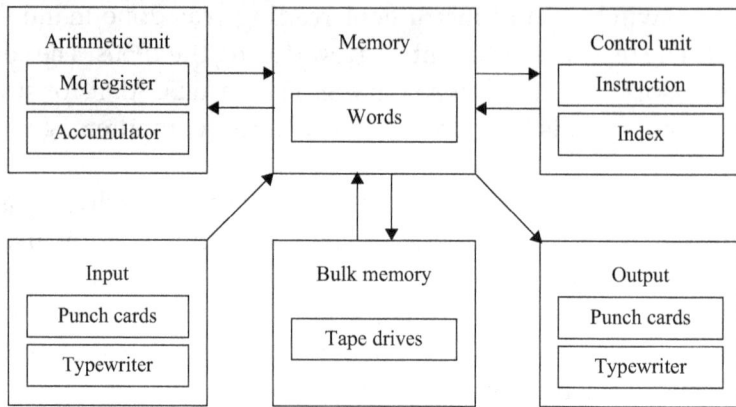

Figure 3.4. Early computer architecture.

What made the stored program digital computer revolutionary was that it was possible for inventors to build things without all the complications of designing and building hardware. All a designer had to know was how to program the computer. Well, that and the fact that mathematical calculations and simulations that couldn't be done by a human in a hundred years now appeared to be possible. There was new interest in mathematics and the sciences.

As the price of computers came down and their capability increased, they began to find use in process control. Initially there was a question of trust, because reliability was an issue. There were a large number of parts in a computer, some of them moving. Vacuum tube computers were out of the question. Early transistorized computers became available in the late fifties.

Again, the number of companies that sprang up to meet the demand was large. Again, time and acquisitions have removed most of them from the market.

We have seen how early inventions first mechanized processes that had been done by people for ages and then regulated and sequenced the mechanized processes. Examples are the weaver's loom followed by the Jacquard and Toyoda looms. So far in this story, the computer has merely given us another way to regulate and sequence processes, and tabulate the results.

Direct Digital Control

Dr. Thomas M. Stout writes about the Ramo-Wooldridge RW-300 computer in "Industrial Computing Control after 25 Years: Micros to Hierarchies." The machine only had a rotating magnetic drum for the storage of about 16K 18 bit words. The processing rate was about 1000 instructions per second. Up to 1024 analog inputs were converted and placed directly into memory, making this a computer designed for manipulating real-time data. Unlike many machines, it also had analog outputs.

The human interface for the RW-300 was a Friden Flexowriter, a teleprinter that could read and punch paper tape. The machine was developed in the thirties and was used on all of the early computers. There were few CRT displays in those days outside of military applications. The system was displayed at the Cleveland ISA show in 1957, with a mock-up of a fractionating tower and a blender.

Tom was deeply involved with the first use of direct digital control (DDC) at Texaco's Port Arthur, Texas, USA, refinery catalytic polymerization unit. The project started in the fall of 1957 and was ready for startup in March of 1959. That time was spent in careful evaluation of what the computer could do, what it was worth, and how it could do it. There was also good communication between the people who would own and operate it and the people who would design and install it.

The Port Arthur system had 103 inputs including 2 analyzers and 14 electrical analog outputs. Five of the outputs went directly to pneumatic converters to set the reactor feed valves. At the time, mistrust of the computer was so deep that most plants would not even consider DDC. The other nine outputs supplied pneumatic setpoints to loop controllers for setpoint control. See Figure 3.5.

After ten weeks of testing, the startup was uneventful except for the reporters from Business Week. The refinery manager duly signed off on the project and the crowds left. Tom wrote, "The finest compliment, however, was paid indirectly by the operating foreman in the afternoon when he requested permission to 'keep the loops closed' throughout the night." Considering that refinery operators are among the most conservative in the industry that was high praise, indeed.

Not everyone praised it, though. The following year a professor at Michigan State University published an article titled "Three Myths of Automation." In it he wrote, "There is now an oil refinery in Texas which is completely controlled

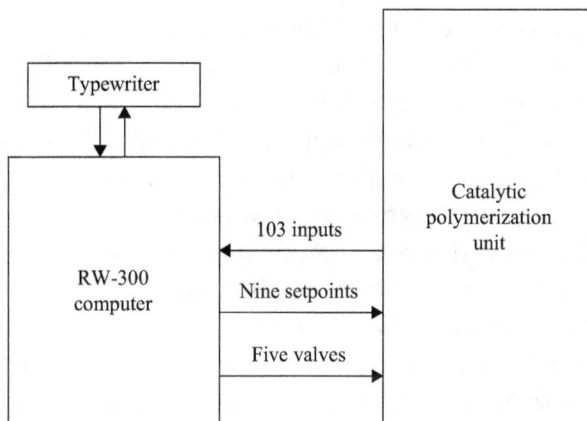

Figure 3.5. Direct digital control.

by a giant computer—and it does a much better job than the human operators whom it replaced." Tom replied that (1) the controlled process is only a small part of the refinery, (2) only some of the process is controlled by the computer, (3) the computer was no giant, (4) the computer did somewhat better than the operators, but not a lot, and (5) no operators were replaced by the computer.

The professor didn't hear Tom's inconvenient facts, but went on to stir up Congress. Seems like there have always been people who are willing to distort a situation to call attention to themselves. At least there was no angry mob that came to destroy the computer, but maybe that's because refinery plant security is pretty good. The refinery was still using the system ten years later when the Smithsonian asked if they could have the computer for an industrial exhibit.

Tom went on to do many more projects, as described in his paper "The Beginnings of Computer Process Control – As Seen by a Beginner." It is one of 26 papers in Industrial Computing Control after 25 Years.

Growth of Computers for Process Control

In 1961, there were about 15 companies making 20 models of computers that could be used for process control. A survey of the petroleum industry showed about ten applications, mostly in refining. The environment for evolution of new companies and applications was restricted by the technology available to build computers—discrete parts, coarse magnetic media with a low bit per pound of device ratio, punch card I/O using large machines, and so on.

Technology improved at a rapid rate, driven by the potential size of the computing market. IBM continued to build mainframes that required fields of magnetic tape drives supported by raised floors for the thick cables that connected all of the devices. That expensive collection of hardware was housed in an air-conditioned room that had at least one window wall to display the computing power in a way that visitors could understand. All of that money required a separate Management Information Systems (MIS) department with top level management. The person that rose to the top was known as the Information Czar, and some of them behaved like royalty.

New companies scaled down the size of computers for process control applications. The work of people like Tom Stout showed that there was a great deal of money to be made by applying high speed computation to large industrial processes. There were also many applications for smaller computers outside of process control. However, there was a limit to how small a computer could be made that used discrete components. Printed circuit boards helped, as did the ability to store more bits per unit volume.

In 1966, there were about 25 companies making 80 models of computers. Some of them were instrument companies, such as Honeywell, Foxboro, Fisher & Porter, and Leeds & Northrup. New companies included Control Data and Digital Equipment Corporation (DEC). The petroleum industry counted 342 applications, most of them in petrochemicals.

The numbers used above came from "The Beginnings of Computer-Based Oil Industry Control Systems," by Gerald L. Farrar (1984).

The environment began to change dramatically with the development of integrated circuits that put miniaturized transistors and components on a silicon substrate and encapsulated them in a 14 pin package.

The Minicomputer Revolution

The idea of building circuits on semiconductor chips was born not long after the transistor, but had to wait for the development of manufacturing technology to build such things. Fairchild and Texas Instruments were early developers of transistors that pushed on into "integrated" circuits. Tools and techniques had to be invented to design circuits on silicon, fabricate silicon circuits, and package them for mounting on printed circuit boards.

The first integrated circuit chips performed simple Boolean functions, such as AND, OR, and NOT. The functions were limited by the number of transistors that could be etched on silicon, which was a problem that yielded to constant development of the manufacturing processes. When 100 transistors per chip became possible with acceptable yields in 1968, the older simple circuits were called Small Scale Integration (SSI). Designers knew that 100 transistors was not the limit, so they called them Medium Scale Integration (MSI).

MSI chips made it possible to build a 16 bit computer on a circuit board of reasonable size, called a minicomputer. Companies like Digital Equipment, Hewlett Packard, and Data General entered that business and thrived on the markets that opened up as the price of computing dropped.

Sadly, there were no standards. There never are when new ground is broken. The leading company (the one with the largest user base) sets a de facto standard for the companies that would compete with it. The irresistible force of change has no chance against the immovable object of a user base.

Each company had its own architecture and instruction set. Programs written for one vendor's machine would not run on another's, even if a standard language like BASIC or FORTRAN was used. The standard languages of the time did not really isolate the programmer from the hardware, partly because programmers were stretching what could be done as fast as manufacturers were adding to their instruction sets.

Pulitzer Prize winning author Tracy Kidder detailed the design of a 32 bit machine by Data General in his book "The Soul of a New Machine," published in 1981. Robert Pirsig, author of "Zen and the Art of Motorcycle Maintenance" said of the book, "All the incredible complexity and chaos and exploitation and loneliness and strange, half-mad beauty of this field are honestly and correctly drawn."

In 1971, there were about 35 companies making over 100 models. New companies included Bailey, Electronic Associates, EMR, Hewlett-Packard, Interdata, Kent, Mitsubishi, Siemens, Taylor, Varian, Xerox, and Yokogawa.

The petroleum industry counted 1,172 applications, most of them still in petrochemicals. Some of the larger process manufacturing companies had in-house engineering groups to apply minicomputers to control applications, such as Dow Chemical, DuPont, Hercules, and Rohm and Haas, to name a few in the United States. Of course the instrument companies developed application engineering groups, as did their sales representatives.

Computer Numeric Control

The first large-scale use of the minicomputer was for discrete processes, as the Computer Numeric Control (CNC) machine.

Removing material from a block of raw material (machining) to match a dimensional specification requires knowledge of the precise location of the material and the cutting edge of the tool. Machinists have used numbers on rulers and positioning handles since the dawn of machining, but World War II caused a shortage of machinists and a high demand for machined parts. Development efforts concentrated on making machine tools automatic.

As usual, development started from what was available. A commercial three-axis milling machine was outfitted with servos that could adjust position in three axes from numbers coded onto punched cards or paper tape. That seems like the Jacquard loom in a new form. The speed at which the tool moves from point to point is also important, as it is limited by the power of the machine and the temperature of the tool tip.

A block of material was put on the milling machine and a paper tape or stack of cards was started in a reader. When all of the tape or stack had been read, a finished part was removed from the machine. This was called Numerical Control. A great deal of work was necessary to translate a drawing into punched holes, but that work wasn't done in a machine shop.

The translation from drawing to tape required lots of computation to move the three axes within the constraints of the machine and the solid boundaries of the part. The first computer that was available to be programmed for this work was MIT's Whirlwind (now laughably slow), demonstrated in 1952. Again, evolution of CNC had to wait for the environment provided by the evolution of the computer, as well as improvements in measuring precise position.

Industrial grade minicomputers in the late sixties enabled many vendors to offer CNC machines. There was a standard language called APT that had been developed for the early machines, but licenses were expensive. The language used to program a Gerber Scientific plotter was adopted for use by minicomputer NC. It became known as G-code and is still in use.

Computers can "crash," as we all know. The crash of a CNC machine is more physical. Machines can either have expensive absolute position measurement or relative measurement, in the same way that pixels can be addressed on a video display. If a relative position machine loses track of its origin, there is no hope of making a useful part. There is also the possibility that the machine may damage itself as the tool holder or part slam into the motion stops on the machine. Even an absolute machine can run a program that crashes the tool into the part. Fortunately, there are ways that a program can be tested for such events in simulated runs, in order to debug the program.

Supervisory Control and Data Acquisition

The supervisory control system for electric power distribution, described earlier in this chapter as SCADA, benefited greatly from the minicomputer.

Improvements in telemetry made it possible to do more interesting things with a computer, and a computer made it easier to code and to decode messages. Algorithms were developed to close the loops for voltage tap and capacitor selection, not with PID but with logic. Measurements could be received frequently, either polled or sent by the substation in a continuous scan. Alarms could be added to the properties of measurements.

The Remote Terminal Unit (RTU) was developed for substations, which directly connected measurements and relays to a box that handled the conversions for telemetry inside a cabinet for wire terminations. Power for the RTU was taken from the 120 volt station battery. A substation still looked like the one shown in Figures 3.4, with the multiplexor replaced by an RTU. A large SCADA system could have several hundred RTU cabinets.

The dispatchers that managed power distribution were used to working with pushbuttons. They, like many other people going through the same thing, did not take an instant liking to the new CRT and keyboard that came with the computer. Acceptance came with graphic terminals that could show the distribution network in layers of detail.

A computer made it possible to do automatic generation control at the generating plants in order to accomplish economic dispatch control. Generation control is done by tweaking the governors of selected turbines to make them produce more or less power. PID control is of no use on a network of hundreds of generators. Control theory says you can only have one integrating controller in a synchronized network. In fact, the turbines have speed governors with 5% proportional band (called droop) that help to prevent loss of synchronization during a load upset, but no tight frequency control.

The complexity of economic dispatch control for a network covering a geographic region containing 1500 generators, 3000 load centers, and 10,000 transmission line segments makes other forms of process control fade into insignificance. This book will not go into great detail on the subject.

It is not practical to keep all of the generators in a region spinning and ready to take loads at a moment's notice. A schedule is prepared of what amount of generation will be needed at what time, based on past history and current events. Generating plants can then start boilers, warm up turbines, and have the generators ready to go when needed. Plants can also know when they can take generators off line and stop burning fuel except to keep things warm. Cold starts take more effort and cost more than keeping a low fire going.

Early attempts to use a computer to schedule a day of producing electricity took more than a day to complete. There was a large incentive to fix that because a few percent improvement in efficiency could save a lot of money for a regional power utility. All of this has to be done in such a way that exactly the same number of cycles of alternating current are generated each day, in order to balance the load among generating plants. Cycles are lost during the day as varying electrical loads go above or below the amount of power being generated. The dispatchers tweak power generation at night to recover the few hundred cycles lost during the day. According to the Air Force during the Cold War, 4:30 AM is the time of minimum human activity.

Initially, telemetry used proprietary signaling systems that were not accessible to the public. The messages were encrypted when modems were used on the public telephone network, and the phone number of an RTU served as a password. Now RTU devices are able to communicate over the Internet, and there are people who are not aware of the security risks they are taking, according to security researchers who locate and gain access to RTU web interfaces with Google.

Programmable Logic Controller

The digital computer was a natural fit for control systems that used Boolean logic in the form of switches and relays that filled cabinets where discrete things were made. Arguably, the ladder diagram used to define the wiring made it possible to build fairly complex systems that an electrician could diagnose without having to learn Boolean algebra.

Figure 3.6 shows a ladder diagram for a typical motor control rung, for motor M92. The vertical lines at each end of the rung are power rails. By convention, the left rail is hot and the right rail is neutral. M92 START is a normally open circuit pushbutton and M92 STOP is a normally closed button. They are located on the operator's panel for stopping and starting M92. The normally closed contact M92 TRIP is on the high-voltage motor starter located in a panel of motor starters in an electrical service room. K92 is a low-voltage relay located in a low-voltage control cabinet somewhere. It has a normally open contact (not shown) that is in the circuit that energizes the motor starter

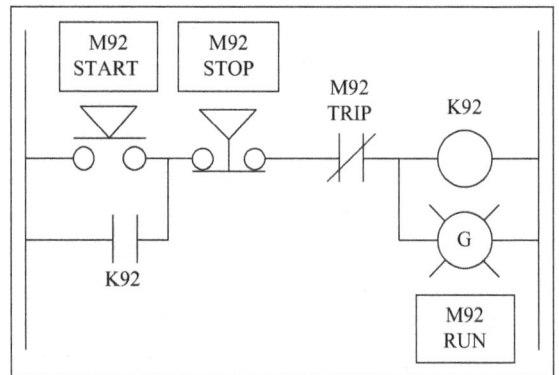

Figure 3.6. Ladder diagram.

for M92. K92 is a normally open contact on relay K92 that closes when relay K92 is energized, sealing the circuit around M92 START so that the relay will remain energized after the start button is released. I92 is a green pilot light on the panel to indicate the state of the motor. Once the start button is pushed and released, K92 will remain energized until the stop button is pushed or the M92 TRIP contact opens when the motor draws too much current.

The Boolean expression is

K92 = (M92 START .OR. K92) .AND NOT. M92 STOP .AND NOT. M92 TRIP

Discrete control rungs typically have a string of interlock contacts to prevent something bad from happening, a contact that momentarily energizes a relay with a seal contact to hold it, and typically a limit switch or timer contact that de-energizes the relay when the required action is complete. Electricians are most often called to find out why something didn't start, which requires finding the ladder diagram that contains the right rung and then locating and testing each interlock to see which one is open. Less often, a relay, timer, switch, or contact has failed. The ladder diagram makes a straight-forward process of elimination possible. If a string of contacts and switches is long, it is best to look for power in the middle and continue searching in the half that doesn't have power.

Electricians have been using ladder drawings since early in the twentieth century. Only mechanical drawings (of parts and machines) are older. There was no such thing as inventing a new method to describe electrical circuits when the computer came along.

The first programmable logic controller (PLC) is credited to Dick Morley, who detailed the design on New Year's Day of 1968. There was more to the design than something to solve simple logic expressions with a one-bit accumulator. It had to be fast so that it could solve the entire ladder in one half of a power cycle, about ten milliseconds. It had to be reliable because if it ever stopped all of the process equipment under its control would stop moving, and not always in a safe state. It had no moving parts, especially not a fan or a disk. Thermal considerations for the components of that time required a rugged case with heat fins.

The PLC was not modular or open. Open systems cannot be completely tested with all combinations of other people's hardware or software. If they can't be tested, reliability cannot be assured. If a PLC fails somewhere in a spectacular way, that is the end for that manufacturer. A sealed case with no fan keeps process contaminants out of the electronics. It also did not have a power switch.

Low energy circuits were avoided in the controller box. In particular, the magnetic core memory used larger cores to require more drive current and produce a stronger readout signal. This greatly reduced the effect of electromagnetic interference. You could wrap coils of electric welding cable around it and weld without disturbing the machine. It also had to handle shock and vibration.

General Motors' (GM) Hydramatic division had been dealing with about 50 feet (15 m) of relay control cabinets for its production line. The controls had to be rewired for every model change. GM engineers knew that great things were being done with solid-state devices. In the summer of 1968, the GM engineers produced a specification for a "standard machine controller" and negotiated a contract with Bedford Associates (Morley's company). Bedford didn't get rich from the contract, at least, not right away. GM bought a few units for testing while they went through the process of convincing themselves that they could replace all of those cabinets with these small boxes. The Bedford system became the famous Modicon 184 PLC.

All that the PLC did was to solve logic with high reliability. There was no operator interface. If you needed one, you built a panel of switches that were inputs to the PLC and indicators lit by PLC outputs. This came naturally to designers who used relays, but computers were beginning to use graphic displays. No PLC vendor jumped into this breach, so a niche was open for new companies. Wonderware introduced the InTouch HMI software in 1987 with some really creative advertizing. The software ran on Microsoft Windows, which was an accomplishment in itself, considering that the word "windoze" was being applied to the early versions. No matter, here was a graphic replacement for all of those lights and switches. Wonderware is now absorbed into Invensys along with the Foxboro Company. Other vendors have similar offerings because they are all trying to solve the same problem with the same technology.

Regulatory Control

The previous systems—CNC, SCADA, and PLC—were not designed to do regulatory control.

Other industries in the late fifties and early sixties were looking cautiously at involving computers in processes that were returning profits. A computer could be used off-line to analyze production results and look for possible improvements. It could be used in-line to examine real-time data from the process and give guidance to operators, who could accept the advice or not. It could be used on-line, as at Port Arthur, with the operator's permission. If the operator didn't understand what the computer was doing, or thought it unsafe, the operator could cut out the computer and revert to backup analog controllers. The operator had the final say on front-line decisions affecting plant safety.

The fifties and sixties mostly consisted of an uneasy truce between operator control and computer control. Few purchasing companies issued a specification that did not allow for backup control if the computer went crazy or died. Tom Stout's achievement is all the more remarkable in that environment. The cost of a computer system favored those processes that didn't impact the plant's profits if they failed but offered an increase in profits if they worked.

Obviously, the cost of the computer system played a large role in that decision. In those days, the capital cost of process equipment outweighed the cost of people. Reduction in staff was seldom considered in the return on investment for a computer project. The existing operators were going to have to take over when the computer failed.

There was also friction between MIS groups whose Czar believed that his empire should contain all computers, and the engineering department functions that wanted to use relatively inexpensive minicomputers (certainly no glass-walled rooms) to do process control. There was a large difference in the timescale of MIS events and process events. The idea of MIS programmers writing process control applications as a very bad deal was obvious even to top corporate management. Sometimes the Czar had his dreams of empire resized, and sometimes a great deal of money was spent proving that it was a bad idea. Some of that persists today, but not in profitable companies.

The following example illustrates the effect of falling computer prices, the effects of installing a computer, and the operator's attitude toward it all.

Wastewater still   A control engineer at Hercules in 1970 became involved with their first computer control project. A proposal was made to the Directors to authorize funds for the project, which was to control a wastewater still in a large operating process for making the raw material for polyester (DMT). It was necessary to keep the project from attracting MIS attention by never mentioning the word "computer." After approval, a computer vendor was selected from DEC, HP, and Honeywell. The boss thought the marketing demos for two of them were too lavish, showing that they had too much overhead. A DEC PDP-11/20 was chosen based on the demo at their factory, an old textile mill in Maynard, MA. DEC's application group said that they could write the control software, so the control engineer spent a week in a small motel describing the control problem to an application engineer, and then went home. Two weeks later, the application engineer said that he hadn't understood a word that had been said.

Another application engineer said that he could write an operating system that would work with FORTRAN code that Hercules would write. The control engineer took a one week course in FORTRAN and proceeded to put together something that would do PID control for the five loops of a distillation column and provide a text interface for the operator. The application engineer endeared himself to the control engineer by the way he handled garbled text input. Instead of the popular "Illegal entry" (do I need a lawyer?) he wrote, "Huh?" It was concise and very effective.

The control system was assembled in an old steel desk with a wooden console for the panel graphic and its five controllers. The desk and the DEC writer terminal were tested and shipped to the plant. The desk took up 10% of the control room space and became the butt of jokes by the operators, who were used to standing at vertical panel boards packed with 3×6 controllers and indicators and 6 × 6 recorders. "Think you've got enough room there for five loops?" The main operating panel board had about 300 loops.

After some training and the uneasiness with something new subsided, the system was tried on the real still. This column was mostly stripping section because its job was to remove methanol from the water that could not be recycled. It was about five meters in diameter and ten meters high, with a smaller diameter for the upper rectifying section. It had a large reboiler. Control was unexpectedly poor, causing great consternation among the engineers. What had happened was that the computer had discovered a problem in the still. That sort of thing was common with new computer installations.

An inspection port in the stripping section was opened and it was discovered that there were no trays in that part of the still. Further inspection revealed that the trays had been driven to the top of the section by excessive vapor from the oversize reboiler at some time in the past. Once the damage was repaired the control system worked quite well. It was never intended to leave the system in place so it was shipped back to the engineering lab. The operators were happy to get the space back. The trial did lead to several computer control systems.

The Microcomputer Revolution

The idea of building computers on semiconductor chips had to wait for the development of manufacturing technology that could build thousands of transistors on a chip, called Large Scale Integration (LSI). A group at Intel began work on a set of logic chips for a hand-held calculator. They concluded that a limited capability computer would be less work than designing dedicated function chips. It took something with the sales volume of the calculator business to justify the cost of designing new chips.

The result was the Intel 4004, released in 1971. One of the developers, Stanly Mazor, said that it was like scaling down an eight-passenger van to a golf cart. It had to be drastically scaled down, because the only package available for chips in 1968 had just 14 pins. The 4004 certainly was not a competitor for the minicomputers of the day, but it was a great advancement for the LSI world.

Improvements followed as customers sought more computing power, and the 8080 was released in 1974. This chip had enough power that it needed an operating system to help programmers make use of it. Gary Kildall of Digital Research created CP/M, which gained wide acceptance. Kildall is most famous, though, for missing a meeting with IBM people that could have made CP/M the operating system for the PC. The stumbling block was a non-disclosure agreement required by IBM, but history does not record why negotiations did not proceed.

Motorola entered the field with its 6800 microprocessor in 1974. This was no copy of the 8080 architecture, though. It drew on the design of the popular

DEC PDP-11 for its architecture and instruction set. Not only was the instruction set different from the 8080, the order of bytes in a word was reversed. The 8080 stored the low order byte first (little endian) while the 6800 stored the high order byte first (big endian). Mazor says he regrets that decision for the 8008, which was made to be compatible with one of Intel's customers. The situation persists today, requiring conditional compiler statements in code written for several different microprocessors.

Intel developed the 8085 in 1976 to meet competition, but it had to be backward compatible with the 8008 and 8080. The chip contained about 6500 transistors built with new N-MOS technology. Development of the 8086 started two years later. It was a 16 bit processor with about 30,000 transistors. Another version, the 8088 had an 8 bit data bus for compatibility with existing memory chips. This was the processor chosen by IBM for the PC released in 1981.

Distributed Control

Microcomputer chips enabled control companies to make their own special-purpose computers. The first major vendor to do so was Honeywell, working with Exxon Research and Engineering to develop a control system that was highly reliable. The first requirement for increased reliability was to end the dependence on a single minicomputer for hundreds of control loops. This led to the use of microcomputers to build control systems for a smaller number of loops and to distribute these systems to locations closer to the process than the central control room.

By this time, video displays were already being used for process control. Exxon had been doing its own research on the operator interface and had some requirements for that. Operators are responsible for too many loops to have to check each loop individually. An overview display was designed that showed the deviation between process value (PV) and its set point (SP) as a bar above or below a line for zero. The bars were presented in groups of eight with 32 groups on the display. Each group of eight had a short name but no instrument tags were shown. The operator could quickly see a that group was in trouble, select it and get a detail of the eight loops, then select a particular controller detail display for adjustment.

Honeywell released the system in 1975 as TDC-2000, a total distributed control system (DCS). It would have been more correct to call it a totally digital distributed control system, but marketing people don't think like that. The initial release consisted of a basic controller, process interface unit, and basic operator station, connected by a data "hiway." Each basic controller had 16 analog inputs and 8 analog outputs, enough to control a unit process like a distillation column. The eight outputs were divided between two cards, to limit the consequences of a single card failure. The process interface unit was a multiplexor that accepted up to 256 input signals of various levels (64 per rack) but had no outputs. The basic operator station displayed the contents of a basic controller or process interface unit with the overview described above. This allowed the operator to do anything with a controller that would have been possible on an upright control panel. The operator station had an optional printer and video copier that allowed screen captures to be saved on paper.

Other control companies introduced their own versions of distributed control using the same technology to solve the same problems, but with marketing and engineering departments developing all sorts of "product differentiators." The result was similar to what you find when you rent a car—it has the right controls, but with different handles and not where you expect to find them.

A distributed control system cuts down on "home-run" cables from devices to the central control room. Wire isn't all that expensive compared to process hardware but the labor for installing and terminating it can add up. Exxon used to bury home-run cables (along with spares) in trenches that were then filled with red concrete. If anybody digging hit red concrete, they had to turn around and dig in a different direction.

Control in the Field   The measurement device in a control loop consists of a physical property transducer called a sensor and a means to amplify, linearize, and convert the transducer output to a standard signal level. The package containing these components is called a transmitter, although the transducer may not be included in the package.

Transducers required compensation for environmental changes, principally temperature and pressure. Analog compensation could be expensive, requiring selected parts chosen after running the transmitter through its rated environmental range. Microprocessors, when their power requirements got below 50 milliwatts, could be programmed to do the compensation using automatic equipment. This made 1/4 and 1/10 percent accuracy possible.

As microprocessors grew more powerful per milliwatt, it became possible to communicate with a transmitter from the control computer. The first "smart" transmitters appeared in the early eighties. At first, communication was used to inquire about a transmitter's health and to adjust its span and zero. This saved instrument technicians from climbing ladders to reach almost inaccessible transmitters whenever an operator doubted the accuracy of a reading, which reduced process operating costs.

The ISA re-opened SP-50 in 1985 to reach agreement on a standard for digital communication with field devices, but there were too many competing methods already established. The work was divided among three of the layers in the OSI seven-layer model: physical, data link, and application. Richard Lasher of Exxon Research and Engineering asked for and got a forth layer called the user layer. His motive was to define control functions that could be run in a field device so that the application layer would have specific goals. He also realized that the microprocessor in a field device had to be highly reliable, so he could add control functions to get inexpensive control loops that could "keep on PID-ing" when communications failed.

The physical layer was the first to complete its work because there's not much to argue about there. Tom Phinney, communications genius, led the data link layer work through staunch opposition from other protocols. Udo Dobricht of Siemens' Profibus led the application layer to design a useful layer. Lasher led the user layer to design control functions with particular attention to their initialization.

The SP-50 work was pre-empted in 1992 by a consortium of companies that were impatient with the struggles in SP-50. This group turned work by Phinney, Dobricht, and Lee Neitzel into Foundation Fieldbus. Lasher's work was rejected as too complex, but there were undertones of Not Invented Here. New work drew heavily on the user layer work for control functions but not initialization, which was done by back calculation. That work became as complex as Lasher's, of course. Meanwhile, SP-50 split into eight communication methods, Foundation Fieldbus among them.

Intelligent Control

The "Handbook of Intelligent Control" (White, Sofge, 1992) is about neural, fuzzy, and adaptive control, which had been available for at least ten years before the book was written. Neural and fuzzy systems are not widely used for industrial process control, although fuzzy control may be used where the controlled variable cannot be measured, such as cement kiln combustion. The concepts still exist and may find their way into future control systems. Adaptive control is now known as model predictive control.

Neurocontrol is the use of neural networks (real or artificial) to produce control outputs. There were several versions of it in the flurry of papers published after 1988. Supervised control uses the desired control signals to train the neural network, similar to Pavlov's work with dogs. Direct inverse control uses movement of a system such as a robot arm to train the neural network with a series of inputs and matching outputs. Neural adaptive control uses neural networks to determine the coefficients of adaptive controllers, indirectly controlling the system. Back propagation of utility adds some measure of performance as a result of the previous calculations to the inputs to the neural network. "Adaptive critics" are an example of dynamic programming that measures more than performance to one goal. The adaptations compensate for changes in noisy, non-linear systems.

It is one thing to express those methods in words and block diagrams, and quite another to express them in detailed equations to be solved. Training involves learning, which requires memory. Some systems were successfully implemented, others foundered on the rocks of inadequate memory and computing speed.

Fuzzy logic is a way of using rules that approximately express various system behaviors in solvable equations that transform inputs into useful control outputs. Logic is usually binary, but fuzzy language allows a range of values between true and false. Neuro-fuzzy control uses neural nets to determine the rules used by a fuzzy logic controller. The neural nets use back propagation or adaptive critics to compare inputs to system behavior and so "learn" by accumulating experience. The trick is to maintain stable control over the range of input values in the presence of noise and time-varying nonlinearities such as wear of a tool's cutting edge.

The cerebellar model articulation controller (CMAC) was developed for robot manipulators, specifically the coordinated motion of a set of joints. Mammals evolved the cerebellum which sits between the cerebrum and the brain stem and spinal cord. A stroke in the cerebellum causes ataxia in some set of joints, which can leave a person able to balance and walk but unable to coordinate the movements required to use a keyboard or bring food to the mouth. The CMAC is a neural net used to cause a multi-jointed robot arm to smoothly move its end effector from one point to another in minimum time. It works well, but requires a large amount of memory.

The Handbook has three chapters on application of neurocontrol to the process industries before it goes on into control of advanced aircraft. We know these techniques as model predictive control. In Chapter 8, "Artificial Neural Networks in Manufacturing and Process Control" by Judy A. Franklin and David A. White, section 8.5.2 Plant Realities, the authors list 14 points. Here are three that will be understood by control engineers who have worked in plants: "We need basic research done by people who are in touch with actual problem sites and needs." "On-line algorithms must converge as quickly as possible." They also observe that operators are comfortable with PID control but less so with neural networks. This is probably an understatement.

Marketing departments in 1992 were beginning to apply words like "smart" and "intelligent" to any device that contained a microprocessor. The authors point out that a few features that help with ease of use are not the same thing as a device that can "comprehend, reason, and learn about processes, disturbances, and operating conditions." An intelligent device should be able to learn from experience and autonomously improve its performance, just as we do. If that could be done, it would be nice to be able to transfer the knowledge to other devices.

Artificial Intelligence (AI) developed a bad reputation because it was ahead of the technology that could implement it. Governments, knowing that they had little intelligence, threw money at AI programs without understanding what was involved. Then as now, this is a bad way to allocate money. Once grant money was available, proposals promising solutions were drafted. The competition was for the money, without regard for facts. Bureaucrats who did not know the facts responded to the bright future described in the proposals and allocated the money. The news media hyped the proposals for AI to replace human intelligence, and companies purporting to solve AI prospered, until the promises could not be met. There was nowhere near enough computing power in the world to make the AI promises reality.

Progress was made, but it didn't trickle down into industrial control. The Defense Department may order a few hundred expensive aircraft with very sophisticated flight control systems and spread the development costs, while an industrial process would have to do development on a plant-by-plant basis, with each one making a profit. Control vendors could spread development costs over similar customers, but they also had to allow for constantly changing technology.

Advanced Process Control

Regulatory control has served industry well by maintaining a small error between a measurement and a target setpoint. The error, however small, means that the controller is always reacting to a change in the process rather than preventing the error. Advanced process control can prevent errors by using some sort of model of the process to predict what has to be done to one or more actuators to eliminate differences between measurements and setpoints. The model is not so good that it can predict fluid flow errors from valve irregularities, so regulatory control (good old PID) is still used for the final control step.

There are degrees of advancement in advanced process control. Simple cascade control may be advanced. Feed-forward control uses a simple lead-lag function to model the process disturbance. These and Smith predictors are examples of advanced regulatory control. There are even ways to decouple two interacting controllers with available DCS function blocks.

Multivariable control requires an accurate model of the process to maintain control. The method is called Model Predictive Control (MPC). Operators have resisted MPC because they don't know what the model is doing. You wouldn't ride in a car with a driver that seemed to be out of control. If the process doesn't look right to an operator, MPC will be turned off and fall back to regulatory (possibly advanced) control. The models may fail outside of a rather narrow range of operating conditions, so the word "robust" was used to describe a model that could handle a wider range of conditions.

MPC use is growing because it can really save money, and vendors are providing better tools for creating and maintaining robust models. MPC was developed for multivariable flow problems in oil refineries, but it can be adapted to a wide range of problems in discrete processes. Discrete MPC still works on flows, such as coordinating machines to get the correct flow of parts to an assembly machine. A simpler example is optimizing traffic flow at an intersection controlled by stop lights. Models may be used to plan optimum paths when there is a selection to be optimized.

MPC has largely replaced fuzzy logic and neural networks for process control, so that new work is not being done on them. There is great interest in neural networks for other fields, such as pattern recognition. It can be expensive to gather the knowledge required to create a model. A process has to be perturbed to see what it will do, which can affect product quality and quantity. Once the model has been built there are no perturbations until something in the process changes enough to adversely affect the model.

References

A Little of Ourselves. The Foxboro Company, 1958.

Industrial Computing Control after 25 Years: Micros to Hierarchies. Proceedings of the Tenth Annual Advanced Control Conference at Purdue, published by Control Engineering magazine, 1984.

Farrar, Gerald L. *The Beginnings of Computer-Based Oil Industry Control Systems*. Proceedings of the Tenth Annual Advanced Control Conference at Purdue, published by Control Engineering magazine, 1984.

Ivanoff, A. *Theoretical foundations of the automatic regulation of temperature*. J. Institute of Fuel, 7, pp. 117–138, 1934.

McCracken, D.D. *Digital Computer Programming*. John Wiley & Sons, Inc. 1957.

Mitereff, S.D. *Principles underlying the rational solution of automatic control problems*. Transactions of the ASME 57, pp. 159–163, 1935.

O'Dwyer, Aidan. *PID Control: The Early Years*. Dublin Institute of Technology, Ireland, 2005.

Russell, J.C. *A Brief History of SCADA/EMS*. At http://scadahistory.com/, 2009.

Savas, E.S. *Computer Control of Industrial Processes*. McGraw-Hill, 1965.

Shinskey, F.G. *Process Control Systems*. The Foxboro Company, McGraw-Hill, 1967.

Stout, Thomas M. *The Beginnings of Computer Process Control – As Seen by a Beginner*. Proceedings of the Tenth Annual Advanced Control Conference at Purdue, published by Control Engineering magazine, 1984.

White, D.A. and Sofge, D.A. *Handbook of Intelligent Control*. Van Nostrand Reinhold, 1992

Ziegler, J.G. and Nichols, N.B. *Optimum settings for automatic controllers*. Transactions of the ASME 64, pp. 759–768, 1942.

CHAPTER 4

Process Operators

The computer provides your hands. I don't think I could work in a conventional mill. This is so much more convenient. You have so much more control without having to go out to the equipment to adjust things.

—Zuboff (1988)

Introduction

Processes need operators because processes cannot see developing problems in themselves, fix those problems before they escalate, or see ways to make improvements. This suggests that operators should be people who can do those things. Operators cannot be hired off the street or out of school and put to work operating a process, because no two processes are alike. This chapter talks about requirements for operators from several perspectives. The operator should be seen as an extension of the process that is necessary for its function as a producer of products and profit. The management extension of the process will be treated in the next chapter.

It seems to be true that a good operator can operate any piece of machinery once the principles of operation are understood and the locations of measurements and handles to manipulate the measurement values have been found. It also seems to be true that good operators enjoy controlling powerful machinery, with a joy that comes from knowing that they do it well.

You experience the need to locate measurements and handles when you rent an unfamiliar car or truck. You know the principles of operation—accelerating, steering, and braking—but it takes a while to find the right handles, even though brake, steering, and accelerator locations are standardized. Where is the turn signal, windshield wiper, light switch, headlight dimmer, starter switch or button, and so on, and what do the plus and minus signs mean on the gearshift? Forget about using a complex audio/navigation system, where the same button does different things depending on when you press it and for how long. If you are lucky, there is a manual underneath the spare tire in the trunk.

Operator Origins

Process operators must be hired at different skill levels for a new process. Managers of established processes prefer to train people from new hires through classroom instruction in the basic layout and functions of the process, and the safety hazards. Even visitors to a plant may require basic safety instruction, such as the meaning of horn or siren signals and what to do when you hear them. This is not common for discrete processes, but there are some that have significant safety hazards.

Once trained, the new hire is paired with an experienced operator to make the rounds of the real process until there is nothing more to be learned—or what should have been learned wasn't, and the process is terminated. The next step is central control room operation, where the candidate is paired with an experienced operator and evaluated. There are advancement steps beyond control room operation as experience is gained on different processes at the plant.

This is the method used to bring new people into the operation of an existing process that determines the profit or loss for the plant. No one is allowed to touch the controls without understanding what each control does. The occasional mistake still happens, but not because the operator didn't know what to do in normal circumstances.

The good operator can see developing problems and fix them before they escalate. An excellent operator sees correctly how to keep problems from happening again. Managers that understand the value of the operator to the continued operation of the process treat their operators well. This was not always so.

Background

There is a very old division between two classes of people—labor and management. The dividing line is manual labor. If you sweat while working, you belong to the labor side. Laborers try to minimize their exertions in the same way that all humans conserve their energy for a time when it is desperately needed. Management tries to get the maximum amount out of laborers.

F. W. Taylor (1856–1915) worked in the machine shops of the time and went on to try to improve the amount of work done per person with efficiency studies and what he called scientific management. Managers should use science (time and motion studies) to learn the best way to divide a work process, and then determine the best tools and working conditions, train the laborers, and give them cards explaining the small part they are to play in the process. This would require enforcement by management. An unidentified machinist responded to Taylor at a debate in 1914:

> We don't want to work as fast as we are able to. We want to work as fast as we think it is comfortable to work. We haven't come into existence for the purpose of seeing how great a task we can perform through a lifetime. We are trying to regulate our work so as to make it auxiliary to our lives.

Prior to the Middle Ages, trading in money was considered contemptible because materialism showed a lack of Faith. Mercantilism also fit under this umbrella. During the middle ages, work as manual labor itself became contemptible. The Patrician class required that a man have passed a year without manual labor before being considered for their ranks. Specifically, in 1241 a town in Flanders excluded from its gentry all criminals, money changers, and people who had done manual labor within the past year. In the Far East at about the same time, the upper classes grew really long fingernails to show that they did no manual work.

In the eighteenth century, cottage workers were deaf to demands for more productivity. The need for more production caused the creation of factories, which moved work out of the cottages, at a time before steam engines made factories necessary. People went to work in a factory as a last resort, and management had trouble with their willingness to work. Fines up to half a day's pay were imposed for being away from your workplace or talking with people not in your own group. Worker turnover was high because there were plenty of other jobs, or you could go back to the farm. Those options are not available today.

Working with the body may not have been appreciated, but it caused a person to use all of their senses to get the job done well, perhaps to keep the foreman from making physical threats. A skilled person had learned how to use the senses to do the job and prevent mistakes. The skills lasted, as do those body skills of bicycling, swimming, and playing a musical instrument. People who couldn't explain how they did something said they had "know-how."

Once a person had personal knowledge (body skills) of how to do a job, it was almost impossible to get them to use better methods. There's a survival advantage to staying with what you know, up to a point—and changing technology kept raising those points. When management introduced a new and faster tool to be used for the job, a person had to adapt or lose the job. The "know-how" was lost and replaced with something else. It is fortunate that humans are adaptable, for the most part.

From Reality to Abstraction

What follows is inspired by Chapter 2 of Shoshana Zuboff's "In the Age of the Smart Machine." Zuboff finished the book in 1988 after hundreds of interviews with people who were making the transition from traditional, manual methods to using computer interface terminals. The eighties were perhaps the peak time for such transitions. In particular, Zuboff interviewed operators and managers in three pulp and paper mills over four years of their transitions. It is an important book for reasons beyond process operators.

Her findings are presented and discussed here because they show human dimensions of operators that don't normally appear in a book on automation. Her observations match my experience with operators during the same period, when Rosemount introduced the RS3 distributed control system in 1982. The first system took operators from overhead chain valves and hose connection panels to group displays on a monochrome 12 inch CRT. Operator training was . . . interesting. The union steward stalked out of the class saying nobody would get him to use something like that. He was back three days later, ready to learn.

Before computers, operators had many ways of using their bodies to achieve precise knowledge. One man judged the condition of paper coming off a dryer stack by the sensitivity of his hair to static electricity near the machine. This provided a feeling of certainty, of knowing "what's going on." An operator described how it felt to be removed from the process and put in a control room:

> It is very different now. ... It is hard to get used to not being out there with the process. I miss it a lot. I miss being able to see it. You can see when the pulp runs over a vat. You know what's happening.
> —Zuboff (1988)

This description is hardly precise data, but it is how an operator expressed his feelings about the change. The operators know what to do, but can't explain why they do that. Perhaps it is more like the physical ability of an athlete.

Sentience is the ability to feel, as opposed to reason, which is the capacity to think. Before computers and the central control room, operators gained sentience for a process as they worked with it. Reason wasn't absent, just mostly informal. Do this if that pipe is too warm, do that if the pulp feels wrong.

These operators relied on their own physical senses to associate changes in the sound, smell, and feel of the process with actions that must be taken to correct what seems to be wrong. Of course, vision is the most important sense of all, with many associations. Operators felt that sitting in front of a computer terminal forcibly removed them from the process, where they could use their senses.

One operator with exceptional verbal skills expressed it this way:

> With computerization I am further away from my job than I have ever been before. I used to listen to the sounds the boiler makes and know just how it was running. I could look at the fire in the furnace and tell by its color how it was burning. . . . I feel uncomfortable being away from those sights and smells. Now I only have numbers to go by. I am scared of that boiler, and I feel I should be closer to it in order to control it. — Zuboff (1988)

Zuboff compares these feelings to going blind, with the attendant loss of control and feelings of frustration. Add to that vulnerability, because you can cause harm by blundering about on the keyboard, which will cause your supervisor to call you bad names or worse. The blindness is temporary, lasting until the operator is able to abstract the process to what is displayed on the computer monitor and learn what to do to affect the process.

Attitudes towards computer control mellowed as people adapted to it. Not all were competent, though. Some told the computer what to do at times, others let themselves be told, even when things were getting bad. Most of those that adapted accepted the computer as a way to do their jobs better with less labor. See the quote at the head of the chapter.

Some operators saw the computer as a risk for the process. If the computer monitors froze or went dark, who would still remember how to operate the process without it? This is still a legitimate question, with the situation growing more serious as more of the process is automated.

Zuboff notes that the operators are undergoing a technological transformation, requiring a new way of sensing the process. Instead of being surrounded by the process, part of it is abstracted to a computer monitor. Instead of feeling a pipe, the temperature is presented as digits on a screen. Instead of physically manipulating something to correct the process, part of the screen is selected and numbers are entered on a keyboard. Today we would liken it to a video game, but then there wasn't much beyond "Space Invaders," which used the then-new Intel 8080.

Trust in a New Machine

In Zuboff, there is a discussion of symbols and trust, illustrated by a story of medieval written contracts. People had been conveying property by spoken word with witnesses. Writing introduced letters that stood for the sound of words, and people didn't trust them. John of Salisbury, a twelfth century scholar, said, "Fundamentally letters are shapes indicating voices. Hence they represent things which they bring to mind through the windows of the eyes. Frequently they speak voicelessly the utterances of the absent."

Operators have to learn new symbols that represent the reality of the plant in order to read computer monitors. This requires time, and may be seen by management as resistance to change. Humans are quite good at symbols, although the ability varies with the individual. The physical plant actually consists of symbols for what is invisibly going on in the tanks and pipes, in the form of pressure gages, thermometers, and level glasses. Discrete processes are considerably more visible. There is a period necessary to transfer trust from a local measurement in the process to a remote reading on the monitor—it takes time to associate meaning to symbols. Training helps, but it isn't usually in the budget. During the early days of a conversion to "glass panels," there are frequent calls to the instrument techs to check that the expensive remote transmitter is calibrated, because it doesn't read the same as the inexpensive gage in the process. This goes away as trust is established, provided the remote readings and the computer are reliable. Another indication of delayed trust comes from the practice of tapping a gage with a needle to see which way it goes, like an old barometer. Early transition operators have been known to tap the monitor screen.

Gaining understanding of the symbols and trust in the readings is one thing, abstracting the process to a series of graphic displays is another. If the process is new, the lead operators are trained in the operation of the process using the flow sheets and piping and instrumentation drawings that the engineers used to define the process for construction. The drawings are not new to them, and training has explained why they have the lines, shapes, and symbols on them that they do.

Sometimes questions from the operators during training can cause enhancements to the design. Few engineers started as operators so operators do find things to change, but the design is congealed if not frozen when training takes place. Enlightened management may teach the operators how to build computer graphic displays, or at least lay them out. Graphics are built as construction nears Mechanical Completion to capture "as built" changes.

Testing and modification occur during startup of the process, beginning with water runs. There will be differences between design operation and water runs because the density and viscosity of water may be far from the process fluids. Water runs further test that the sensors and valves are in the right locations and functional, even as the water flushes out drawings, tools, and lunchboxes.

One of the managers at a mill saw the need for training, and won an appropriation of resources to put together a classroom with a simulator. After many hours of teaching and getting reactions, he told Zuboff he had discovered that:

> They haven't learned to trust the machine to tell them what to do. This trust does not come naturally. It will only come when they really understand how it works.

Also:

> One of the real stumbling blocks in learning to trust is that if you are in a room with a screen, you have a hard time convincing yourself that something is really happening.   —Zuboff (1988)

This last quote is why a major consideration in HMI and control system design is the response time to an operator action. One second is about the limit; otherwise the operator is likely to issue the command again. This is not harmless if the command is to toggle a switch or raise a setpoint. Major software vendors use a "progress bar" if something will take a while, but that feature is not present in most control systems.

Zuboff relates a story of a discrete process shop foreman relocated to a computer monitor, as told by a vendor's employee who had worked on the job. The foreman had lost his sense of what was going on down on the floor. He had to continually send someone down to the floor to check because he was losing the feel for what he was doing, even though he could call up anything he wanted in the data. The vendor's response was to build in some redundancy so that things could be double-checked automatically. It was a technical solution that was irrelevant to the problem since the foreman was sophisticated enough to know the potential technical solutions. His problem was that he wanted to feel it, be involved with it, and he had lost that. The vendor said that we tend to dismiss this as defensiveness, but there may be more to it than that.

There is more to it, and it relates to trust. It takes time and understanding to develop trust, as anyone who has entered a relationship with intent to mate knows well. If management doesn't understand this need to develop trust, things can go very badly. If management understands this as a change that requires change management, with soothing words about how good it will be and training to make the unfamiliar at least understandable if not familiar, then the transition may be successful. Training should aid the operators to transfer the physical sensations of the process to the symbols that are available on the monitor. The ability to know when something is wrong and the ability to do something about it should follow. This doesn't work for everybody, as this operator says:

> If I had to sit at the computer for twelve hours . . . I don't know, it's horrible thinking about it. When you walk around, you see things that make you think and ask questions. If you just sit back and look at the numbers all the time, you become a machine like the computer.
> —Zuboff (1988)

Back to the present

That was then, and this is now. People haven't changed, but successful operators have been able to abstract the physical process to displays on a computer monitor, and to use a keyboard (not necessarily QWERTY) to make changes in the process.

As an aside, the QWERTY keyboard illustrates how a large user base can prevent change. Christopher Sholes invented the typewriter in 1866. The keyboard was open to any design, but Sholes chose one that kept adjacent characters striking the paper from being so close as to cause the letter keys to jam. The design has nothing to do with efficient typing, but it became the standard when Remington adopted the present layout in 1878. August Dvorak and William Dealey studied the way people typed for twenty years, finally developing a more efficient keyboard in 1932. People trained to use the keyboard were two to three times faster than standard typists and had less repetitive stress injury, but the user base of people and typewriters was too large to even consider retraining. The Dvorak layout became an ANSI standard in 1982, but it is not taught in schools. Programmers like the layout, partly because the letter O is nowhere near the number 0. The faster, more efficient Dvorak keyboard is still a curiosity except among those who have taught themselves to use it.

The following job description and testing requirements should give a better idea of what today's operator faces.

Job Description

The following information is a condensation of job descriptions and requirements found on the Internet in August, 2012. The list is for fluid process jobs, where the fluids are contained in vessels and pipes to keep them from leaking away into the environment, or to maintain pressure. The list for discrete process jobs, where things hold their shape without pipes and vessels, will be similar except for specific machine operating skills.

The successful candidate must appear with the following attributes:

Possible citizenship requirement and background check.

Have a valid driver's license (to drive company vehicles on errands).

Education: High school diploma or equivalent, or relevant military education. Prefer additional technical school training. (Or not—some employers don't want to untrain new employees.)

Experience: (List varies with the job level, may allow military experience.)

Personality: Work well with others or alone, take direction (and give direction for higher level positions), prefer neatness to disorder, safety conscious, obey laws and plant rules, have a proven work ethic, open to change, work as a team member when required, able to resolve problems without showing anger.

Aptitudes: Mechanical, fault diagnosis, verbal and written communication, use of computers, use of job-specific equipment, ability to see a diagram as a map of the process, use of hand tools (may include specific tools), use of test equipment (may include specific equipment), use of two-way radio. (Some or all of these may be tested before an offer is made. Employers must be able to show that a test is able to predict future job performance, if challenged.)

Physical (may be tested): Good health and condition, free of mood-altering drugs, good eyesight and hearing, able to perform specified body motions, able to stand or sit for specified periods, walk for specified miles a day, lift or move specified weight, climb a specified number of stairs in a specified time, climb a ladder, and possibly have no fear of heights or tightly enclosed spaces. (May have to be clean-shaven to wear protective masks; smokers are not welcome in plants with hazardous atmospheres or where smoke breaks are not possible; allergies may need to be discussed.)

The successful candidate must be willing to do the following things:

Work rotating shifts of either eight or twelve hours. (Many prefer four days of twelve hours and four days off before rotating instead of the progressively rotating eight hour shift with one day off per shift change.)

Work overtime as required, or be on call at any time of the day or night.

Wear protective equipment as required, even though the hazard may not be obvious.

Work in extreme weather conditions and otherwise hazardous environments with the proper protection.

Use a steam or water hose to clean up a spill.

The successful candidate will assume the following responsibilities:

Maintain awareness of safety issues and requirements.

Report any safety concerns immediately to the supervisor.

Don't let anybody touch the operator's console that hasn't asked for your permission.

If someone else operates the process, you are responsible for what they do.

Maintain safe operation of the process within specified limits.

Maintain shift events log, personnel in the area log, and others as specified.

Perform routine checks of equipment, or communicate with field operators for that purpose.

Maintain area equipment to ensure operational reliability.

Make necessary adjustments to maintain normal operation of equipment.

Prepare equipment for maintenance activities when maintenance is required.

Comply with lock out, hot work, and other safety procedures.

Assure that maintenance is complete before resuming operation.

Perform housekeeping in the process area to ensure a safe and clean workplace.

Report any unusual situations to the supervisor as they occur.

Perform additional duties as assigned (a very wide-open requirement).

Testing

Possible pre-employment tests that may be required:

Bennett mechanical aptitude—The Bennett Mechanical Aptitude Test has been used for 60 years to measure a person's understanding and comprehension of spatial perception, reasoning, and his or her general aptitude in the areas of mechanics. The test is listed among the requirements for the job by many people applying for a job as a process operator. See http://us.talentlens.com/bennett-mechanical-comprehension-test for details and a link to a manual for testing. There are no samples but you can pay to take the test on the web. Search for the test by name to find several other sites.

COBRA—The Console Operator Basic Requirements Assessment is a four hour test at a simulated operator's console with increasingly difficult operating problems. It has been used since 1993 for about 50,000 applicants and has become established as a statistically significant indicator of performance. See www.getcobra.com. From that website:

COBRA was developed to assess the most important cognitive abilities needed by individuals who work a console position. For example:

Selective attention (the ability to concentrate in the face of distractions).

Problem sensitivity (the ability to determine at the earliest stage when something is wrong).

Time-sharing (the ability to multi-task).

Reasoning (the ability to think through how an action will trip off a string of subsequent actions).

Resistance to premature judgment (the ability to not "over-operate" the console).

Speed of closure (the ability to make sense out of large amounts of process information).

Response orientation (the ability to quickly choose between two or more actions).

POSS—The Plant Operator Selection System is a set of three batteries of tests originally designed for electric power generation plant operators. A typical battery of tests contains:

Reading Comprehension tests the ability to read and understand equipment manuals.

Mechanical Concepts tests the ability to understand mechanical principles typically found in a process.

Mathematical Usage tests the ability to convert engineering units, do problems in algebra, and solve word problems.

Spatial Ability measures the ability to visualize the final assembly of a set of objects.

Tables and Graphs measures speed and accuracy in extracting data from tables and graphs.

Search the web for "POSS test" to find current information on the tests, including tutorials for taking the tests.

The following is a list of tests available from www.psychometric-success. com with sample question and answers:

Abstract reasoning—tests the ability to complete a sequence of four diagrams. (It is considered a culturally fair intelligence test because it does not use words or numbers in the problem statement.)

Concentration—tests the ability to eliminate false patterns, as a timed sequence of increasing difficulty.

Fault diagnosis—tests the ability to learn arbitrary functions and detect the fault in a sequence.

Mechanical reasoning—thorough test of math and reasoning, not just gears.

Numeric reasoning—missing numbers and word problems, calculators allowed.

Spatial Ability—tests the ability to mentally match shapes when one of them has been rotated and find the shape that could be assembled from a set of smaller shapes.

The Actual Job of Operating

There are many descriptions of a process operator's job, and many of them miss the point.

An operator is trained to understand and run the process and then left alone with it. There are supervisors and technical people available to help, but the operator is first responsible for the safety of personnel and equipment, then quality of product, then quantity of product. Operators work in shifts so that there is time to unwind from the tensions of responsibility.

Operators include machinists in the discrete manufacturing processes. A machinist is one who knows how to safely operate a machine to produce quality parts at the best possible rate of production. In this sense, a machine is a tool that requires skill to operate, so a metal lathe rather than a punch press (unless the operator is responsible for changing dies and setting up the machine for quality work).

Actually, there are no lathes in mass production operations, just computer numerically controlled (CNC) metal shaping machines. The machines are completely automated, doing their work behind bullet-proof glass where no human can touch it. There is the problem of removing finished parts and loading blank parts, but that is what one-armed robots do all day. The machine's computer is programmed by a human skilled at reading mechanical drawings and converting them to machine codes, or a computer-aided design (CAD) file is converted to CNC settings. A human is only required when a broken tool or part challenges the bullet-proof glass shield.

An operator is responsible for quality in so far as quality can be measured and displayed in real time. You can't manipulate what you can't sense.

Being responsible for safety means more than just the rules for precautions, such as don't stand under a load carried by a crane. An operator must know what can go wrong, and the warning signs for impending failure. Years ago, nitroglycerin was made in 500 pound wooden vats containing strong nitric and sulfuric acids with the glycerin added by the operator, who also had a large wooden paddle to stir the mixture.

The vat was surrounded by a catwalk where the operator walked, and the whole was surrounded by thick dirt walls that would direct the explosion upwards if things went wrong. The dirt walls were pierced by tunnels from the catwalk to the ground. If the operator saw brown or white vapor rising from a hot spot that couldn't be cured with the paddle, he (history records no women taking this job) dived down a tunnel and hoped to live through it.

No one makes nitroglycerin that way now. Today's operators have to infer thermal runaway from temperature or pressure measurements. The operator is safe inside a concrete blockhouse with thick walls while trying to stop the runaway or pressing a button to kill the reaction and ruin the product, but saving equipment or lives.

Operating Environment

The central control room operator works in an environment that is superficially similar to an office area except that there are no cubicles or other barriers to human communication. Lighting is subdued to make the monitor screens more visible. The desk is wide to accommodate as many monitors as needed to communicate the state of the process to the operator. Some of the monitors are dedicated to specific information; the rest may be changed by the operator to focus on whatever is of concern at the moment.

Operators are sensitive to the position of things that they use, such that moving something out of its normal position is disorienting. One of the early distributed control systems had "soft keys" below the monitor screen. Different displays provided different labels for the keys (actually push buttons). Operators hated them, in spite of the marketing hype. Each button had to have the same meaning at all times. The control industry learned its lesson and moved on. It had seemed like a good idea at the time.

The monitor displays provide process state information. The operator affects the process with keyboards and radios. A QWERTY keyboard is provided for the rare instances when characters must be entered in data fields. An operator does not have to know how to touch-type. Each control vendor has a different version (like the auto industry) of a keyboard with specific operating functions. There are configurable keys to select process unit displays. After a display is chosen, a positioning device such as a mouse or trackball is used to select a control point on the display. A single SELECT key provides a detail display of the abstract object selected. If it is a controller, special keys are used to change the controller setpoint, mode, and if applicable, output. These are the same options that were available to the ancient operators who stood at control boards full of controllers, indicators, and trend recorders.

When an operator has to diagnose a problem, a trend display of the previous values of process variables is essential. In the days before computers, physical recorders showed the trends in red, green, and perhaps blue lines on charts that moved at 2 cm (3/4 inch) per hour. The pens could not overlap, so there was a time offset depending on the color. Worse, the time of day on the chart depended on the wristwatch of the operator that last changed the chart paper supply, or the ink for a pen may have run out. The computer fixed all that, if the system's computers were all synchronized in time. An operator could look at a trend display and see what happened first, thereby isolating the cause of a problem.

An example of trouble shooting with recorders was the case of a faulty compressor surge control system. A large axial compressor in a nitric acid process was tripping more often than usual, and surging with a BANG that threatened to tear it loose from its mounting each time. All of the equipment tested as functional, so a high speed (several centimeters per second) recorder was attached to signals in the surge control loop. It revealed that the surge valve had opened properly, so the search turned to the vent piping, which went up to a vent silencer high in the air. The line was long enough to need an expansion joint. A crane was enlisted to remove the silencer and lower it to the ground, where the liner from the expansion joint was found plugging the flow. Fixing that stopped the surges. The false trips were coming from an open junction box under a steel grate stairway where rain water could collect, and be made conductive by the small amount of nitric oxide in the air around the process.

One or more displays may be dedicated to the status of process alarms. Much more will be said about this under the heading of Alarms.

If the process has field operators, the control room operator has a radio that communicates directly with the field. Cell phones are inadequate because the connection has to be dialed and made, and signal levels are low amid the steelwork of the process. Local radio is much better. Another radio may be used to talk to other operating departments, like the barge dock, the warehouse, and maintenance. The last resort is the telephone, which can reach anyone, for example, the engineer who last made a change to the control system, at the fabled hour of 3 AM.

The operator still processes paper as forms and records, so the desk has traditional things such as pens and pencils, staples and tape, three-ring binders with log sheets for work done by others on the process, and clip-boards for sign-in sheets and records of people entering and leaving the process area, in case an accident renders one of them unidentifiable.

A process may have two of these operating positions for redundancy, and for times when two operators are required to get the process back under control or shut down quickly. The second position may also be used by an engineer called in to diagnose a process control problem. The engineer may look at everything, but must ask the operator for permission to change anything. The operator really does represent the single point of responsibility for the process to the outside world.

Other control points may include hardware buttons for emergency stops of various kinds (if a shutdown system has not already activated), the status of purge air for the control room to keep process gasses out, and a white-board for notification of different transport routings in the process as well as equipment that is not available. Video monitors may show visual parts of the process, such as the flame at a flare stack or a part of a machine that may jam.

Alarms

An automated control system does not demand constant vigilance by the operator, allowing him or her to "perform other duties as assigned." The control system notifies the operator of an exception by using an alarm system that draws the operator's attention to the alarm. The alarm system sorts notifications from elements of the control system by priority and presents them to the operator, typically as one line reserved for the highest priority alarm on all operator displays. There are also displays of lists of alarms sorted by priority, state, and time. Alarm data is kept in a history, which may be reviewed with a display for that purpose. Selecting an alarm line typically brings up a display containing the control element that generated the alarm.

Alarms are prioritized because they do not all have the same importance. Typical priorities are critical, advisory, and status. There may be divisions within each priority. The critical priority is reserved for alarms that require immediate attention to avoid loss of product or worse. The advisory priority is intended to let the operator know that something in the process is not normal. The status priority is a catch-all for other notifications of problems that the operator may be able to fix. There is no point to notifying the operator of things that do not affect the process and only other people can fix, unless the operator has so few alarms to handle that it is feasible for him or her to relay the problem to another person.

Alarms were expensive in the days before computers, so they were reserved for things that really mattered. These alarms were announced with a loud horn, klaxon, or siren. The first thing an operator reached for was a button to silence the alarm and acknowledge that it had been seen. This also stopped the flashing lights behind a square of translucent plastic with filled engraving that identified the alarm. The horn just adds to the stress, so it was not uncommon for an operator to stuff a sock in it, if not disconnect it. Nobody ever knew who did that, of course. The flashing light was adequate notice.

Control elements in a computer are blocks of data containing the configurable attributes of the element, including alarm settings. Alarms became cheap, and vendors invented new alarms to provide temporary product differentiation. An architect and engineering company working on a time and material cost contract would not fail to take the time to configure every available alarm. The number of alarms that an operator might experience became staggering. In most cases, the loud alarm sounds had to be muted to allow the operator to function.

Rothenberg shows the exponential growth in configured alarms per operator station in Chapter 3, Strategy for Alarm Improvement, with Figure 3.2.1 going from a few hundred in 1960 to near 4000 in the year 2000.

They didn't all go off at once, partly because half of them were mutually exclusive (high and low). Alarms went from being manageable to unmanageable. That problem is the subject of Rothenberg's book.

Computer control systems were originally designed for continuous processes because that's where the money was. A continuous process runs in the same process state for long periods of time. Alarms call attention to deviations from the normal state. To oversimplify, alarm management becomes an exercise in justifying the existence and priority of the alarms, balancing the use of an alarm with an operator's ability to act effectively to correct it.

Batch and discrete processes routinely change states as they perform a sequence of operations required to make a product. What is alarming in one state may not be alarming in another, but commercial alarm systems couldn't handle that. Whatever executed the sequence of operations also had to adjust alarm setpoints as states changed. Alarm management now has to take the states of the process into account as decisions are made about existence and priority.

All process measurements have noise. Fast sensors like flow, pressure, and level are more susceptible to process noise than slower temperature sensors. Noise would make an alarm "bounce" on and off, which annoys the operator. Alarm hysteresis is universally used for analog measurements to establish a distance away from the alarm point before the alarm will turn off. It cannot turn on again until it gets back to the alarm setting. Discrete sensors have only two values, so hysteresis is not possible. A time delay is used instead, either to delay the alarm or to delay clearing the alarm. These are additional configuration settings for each alarm, unless a single percentage hysteresis applies to the high, low, and deviation alarms possible for an analog value.

Rothenberg discusses other considerations in Chapter 4, Alarm Performance. Alarms may be found to be paired, making one of them redundant for the operator. One alarm may cause another, or they may be related because they occur together within five minutes most of the time. Alarms that stay active for more than one shift must not have been alarming after all. Alarms that are not even acknowledged for more than a day have ceased to be alarms. Then there is the nuisance alarm, rather like telemarketers calling incessantly. An alarm becomes a nuisance when it remains in force while it has no meaning, such as low flow for an unused steam line with condensate boiling out of the transmitter's impulse lines. This is a state problem, because the steam line is in a state where a flow measurement is not possible.

There are many things to do to manage process alarms so that the operator is never distracted by an unnecessary alarm. When an alarm occurs, it is like a computer interrupt. The operator must stop the current activity, store its status, and begin thinking about the reasons for the alarm. This may take seconds to minutes, depending on what was going on. After coming up with a reason for the alarm, the operator takes some sort of action that affects the process. The process then takes time to respond to the action before it starts to move in the correct direction. Normally, that ends the incident and the operator can resume the previous activity.

In Chapter 5 Rothenberg describes an important concept for dealing with situations that get out of hand after the process fails to respond to operator action. The concept is "permission to operate." Normally the operator is the one to give others permission to operate, but when the operator is failing to correct the problem, operation is no longer possible. The operator's permission to operate is withdrawn and efforts are concentrated on bringing the process to a safe state or shutting it down entirely. The operator, after all, is focused on trying to make product. Management must have some way to change that focus when the situation cannot be controlled. Rothenberg goes on to discuss ways of doing this that make fascinating reading, but are outside the scope of this book.

The International Society for Automation revived the SP-18 Instrument Signals and Alarms committee as ISA18 in 2003. They have done great work with some very talented people. The result is available as ANSI/ISA-18.2-2009, Management of Alarm Systems in the Process Industries, published in 2009. The committee is still active, with the following working groups preparing technical reports that will help users to understand the standard:

WG1 Alarm Philosophy.

WG2 Alarm Identification and Rationalization.

WG3 Basic Alarm Design.

WG4 Enhanced and Advanced Alarm Methods.

WG5 Alarm Monitoring, Assessment, and Audit.

WG6 Alarm Design for Batch and Discrete Processes.

Accidents

The operator can interact with other operators and the people who have business in the operating area, as well as visits by supervision and engineers who want to try something. Humans are social beings, for the most part, and so this contact is necessary to relieve the tension that can build up while waiting for the unexpected to happen. Reducing the number of operators for a process leaves fewer opportunities for relieving tension. This can lead to accidents.

Industrial processes need protection from the consequences of accidents, including loss of product, loss of production time, loss of equipment, and loss of life. The probabilities and consequences of loss are different for discrete, batch, and continuous processes, but nowhere are they zero. They are highest for processes with energetic or poisonous chemical reactions.

Here are five layers of protection that are commonly associated with processes:

- Process design—anticipate the hazards and design to contain the consequences of things that can go wrong, because they will.

- Process control—use automatic control where possible to hold process conditions at design values. Use an alarm system to notify an operator that control has failed when it does.

- Process operator—the human ability to detect changes in patterns, analyze the cause of the change, and create a way to deal with it is estimated to have prevented loss 80% of the time.

- Shutdown system—if bad things can happen faster than a human can respond, use automatic controls to bring the process to a safe state when it gets too close to the physical constraints of process equipment or an explosive mixture of materials.

- Mechanical safety system—for example: safety valve, rupture disk, break-away tool holder, or shear pin.

Four of the five layers are difficult to change after the process is built. Only an operator can deal with unanticipated events or more likely, combinations of events. Studies of major accidents show that it is never just one thing that causes a major accident (see Kletz and Perrow).

Not all people called "operators" are capable of saving a process from unforeseen events. Some of them really are just there to carry materials or push buttons, but that's an expensive way to run a high-volume process. Not all of them will learn on the job and become better at it. Petrochemical operators are among the very best, because their processes generate the most profit—and management knows what they are worth.

Perrow's book "Normal Accidents" (Perrow, 1999) describes four levels of a system (process):

- Component—the smallest part of a system that may cause an accident (valve).

- Unit—a group of components that perform a minor process function (steam generator).

- Subsystem—a group of units that perform a major process function (steam generator and condensate return system).

- System—the process (electric power generation).

He uses the term "incident" for a problem with a component or unit and "accident" for a problem that affects a subsystem or system. Then he says, "The transition between incidents and accidents is the nexus where most of the engineered safety features (ESFs) come into play ...." An ESF can be one of the components that fail. Further, "Humans are a part of all systems considered in this book ...." as components or subsystems.

Humans that are part of a system that had an accident are very likely to be blamed for the accident. Sadly, it is human nature to blame others for mistakes that we have made. It is even more likely if the person being blamed is not alive to present a defense.

Kletz ends his book "Learning from Accidents" (Kletz, 2001) with an analysis of the term "human error." He divides the causes of human error into four groups:

- Mistakes—a human had the wrong reaction to an event, usually because of inadequate training.

- Violations—a human knew what to do, but did something else. "If the safe way of doing a job is difficult, people will find an unsafe way."

- Mismatches—between a job and the person hired to do it. This needs to be detected before something happens.

- Attention lapse—a person knew what to do, intended to do it, but failed to do it; often the result of stress or distraction. This needs to be detected and the work situation changed.

Note that all of those causes can be addressed by management, and yet it is often management that blames the accident on operator error. Blame does not solve the problem—action is required. But deeply ingrained human nature demands a scapegoat—an animal to be driven into the wilderness carrying the sins of the people involved.

Consider the case of a very unfortunate operator in the generator room of a large paper mill with a cogeneration plant. Such plants used a number of turbines and generators to allow a high pressure, high efficiency boiler to reduce steam pressure to levels required by the plant. One day, the hydraulic governor for a turbine failed in such a way as to apply full steam flow to the turbine, as well as spraying oil over the area. This was a very rare and unexpected fault. The manual shutoff valve for steam to the turbine was in the spray, so the operator couldn't get to the valve. He did what he thought was the next best thing and tripped the breaker for the generator. The breaker might have tripped anyway, but it was the wrong thing to do. The unloaded generator was spinning up towards infinite RPM when it flew apart, throwing large chunks of metal everywhere and out through the roof. Miraculously, no one was injured or killed. Reclosing the breaker would not have worked since the generator was no longer synchronized to the power bus.

Was the rattled operator at fault for doing what he thought was the right thing? The management that didn't train him thought so, and fired him. An alternative would have been to retain an employee who had learned a lesson he'd never forget. Unless, of course, there had been too many near misses.

Retraining Displaced Employees

The following is controversial, because it contradicts the belief of many managers that employees displaced by automation can be simply retrained to do work required to support automation. The facts do not support this simple transfer of blame for causing the loss of one or more jobs back onto the affected employees.

Certainly there are people who can be retrained to a higher level of skill than what they were using. Some lacked the motivation to advance, but have it now that the job is gone. Some were afraid to try to advance, having had no encouragement from management.

It is possible for the brain to lack the structures or connections required to build new skills. Some people are born with perfect pitch and other musical talents, while others can develop musical talents when they are young. Beyond an age before puberty, it is no longer possible to develop perfect pitch or become anything other than tone-deaf. Cats blindfolded until three weeks old behave as though blind for the rest of their lives because the necessary brain structures didn't develop or connect.

Studies of psychopaths (about 1% of the population) show that they have no ability to read the feelings of other people in facial expression or tone of voice. This makes them efficient predators, while the rest of us can be deterred from attacking by the victim's behavior. Brain scans of psychopaths show no activity in ten regions of the brain associated with empathy. See Robert Hare's "Without Conscience: The Disturbing World of the Psychopaths among Us" or "Snakes in Suits: When Psychopaths Go to Work" for more information.

Similarly, but with a great deal more controversy, the ability to reason from observation to cause depends on the development and connection of brain structures at an early age, if the genes have provided those structures. Intelligence tests are designed to test memory and reasoning. You can't reason if you don't have a long-term memory for facts and a short-term memory for assembling the facts that support a cause, while rejecting memories that are not relevant. You also need to be able to concentrate without being easily distracted from your task. Finally, you must be able to effectively communicate your thoughts to others or to the controls of a machine or process.

Consider the grade scale used in testing students, where C is average and both A and F are rare. Motivation is as important as intelligence, in the sense that the A student requires both, but the failing student may only lack motivation, for any number of reasons. There may be physical dyslexia or other coordination problems. The student may be left-handed, and the teacher may try all sorts of things to "correct" that, although there is more tolerance for sinister people now than there was fifty years ago. The subject may conflict with religious beliefs, or the student may believe any of the stereotypes that humans form because they can only remember fragments of another person's personality. It has been shown that people don't store a complete facial image for everyone they know, just the bits that are unique. If the entire face can't be stored, how much less the entire personality?

All of this would suggest that it is not going to be possible to retrain everybody. There are people in academia who believe that intelligence doesn't matter, and that anybody could get a doctorate degree if they would just work harder. This would seem to be a case of modeling other people based on yourself. If you could get a doctorate, anybody could. This is obviously magical thinking (Hutson, 2012), unrelated to reality, and yet some of these people are the teachers who influence how money is spent for education.

References

Hutson, Matthew. *The 7 Laws of Magical Thinking: How Irrational Beliefs Keep Us Happy, Healthy, and Sane.* Hudson Street Press, 2012.

Kletz, Trevor. *Learning from Accidents.* Gulf Professional Publishing, 2001.

Perrow, Charles. *Normal Accidents.* Princeton University Press, 1999, first published by Basic Books, 1984.

Rothenberg, Douglas H. *Alarm Management for Process Control.* Momentum Press, 2009.

Zuboff, Shoshana. *In the age of the Smart Machine: The Future of Work and Power.* Basic Books ISBN 0-465-03212-5, 1988.

CHAPTER 5

Management

Management is doing things right; leadership is doing the right thing.

—Peter F. Drucker (1909–2005)

If there is any one secret of success, it lies in the ability to get the other person's point of view and see things from that person's angle as well as your own.

—Henry Ford (1863–1947)

Introduction

A manufacturing process, if it makes a profit, contributes wealth to the economy that provides the infrastructure that makes the process possible. A process can't create wealth all by itself. There has to be an organization to handle the money and keep the process going. A wise man once said that a corporation isn't a person; it's a bucket of money that is managed by people in the company. A more general term for corporation is business.

A manufacturing process is part of a business. Businesses all have processes, but not all processes are concerned with manufacturing. Business processes are hierarchies within hierarchies, as shown by typical organization charts. Each hierarchy is headed by a manager, who may have sub-managers and various levels of office or field sales workers.

Just as no two individuals are alike, no two management hierarchies are alike. However, there are identifiable similarities. Various models of businesses suggest organizations of hierarchies. The following is a view from an altitude that renders most of the details invisible, but it describes attributes of an organization that will be found in most manufacturing businesses.

Management Organizations

Discrete Processes

Professor R. Thomas Wright, in his book "Processes of Manufacturing" (1987) devotes one percent of the book to descriptions of manufacturing organizations. This minimalist approach provides a simple introduction to a complex subject. There are five areas of managed activity in Wright's manufacturing corporation. The central area is Production. Surrounding it are Industrial Relations, Marketing, Financial Affairs, and Research & Development. See Figure 5.1.

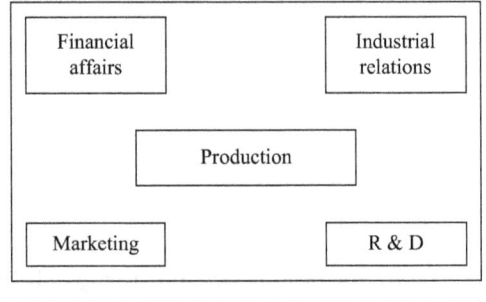

Financial affairs		Industrial relations
	Production	
Marketing		R & D

Figure 5.1. Production and support.

Production is managed to produce scheduled products to specified quality standards, and to engineer improvements to the process. Industrial Relations includes improving relations between management and personnel, labor unions, and the general public. Marketing identifies target markets and develops ways to promote, sell, and distribute the products. Marketing also keeps a close watch on competitive activities. Financial Affairs monitors and controls the bucket of money. Research & Development specifies new or improved products and processes, if Marketing thinks they will sell.

Four functions of management are applied within each of the five areas:

Plan specific goals and courses of action to meet them.

Organize people into structures and define job requirements for each person.

Direct and motivate people to do their jobs.

Control management activity by monitoring and reporting progress towards goals.

Figure 5.2 shows the four managed activities of Production, which are listed below:

Production Planning schedules resources (people, materials, and equipment) for products.

Manufacturing uses scheduled resources and energy to produce scheduled products.

Manufacturing Engineering designs and creates projects required for facilities and equipment.

Quality Assurance does what needs to be done to assure adequate product quality.

There are three types of manufacturing, each requiring different management styles:

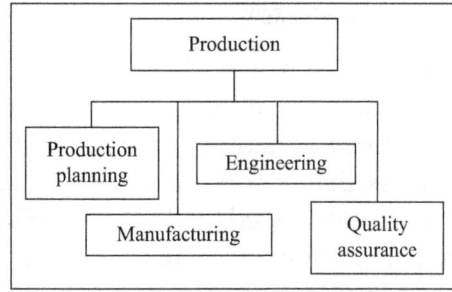

Figure 5.2. Production hierarchy.

Custom manufacturing produces a small quantity of products built from a customer's specifications. Skilled workers, general purpose machines, and a wide variety of material sources are required. Product Planning must work with the customer to produce mutually agreed upon specifications and schedules. Quality Assurance may have to measure every piece. The R&D activity does not design new products, but must keep up with new technologies.

Intermittent manufacturing produces quantities of identical products that are too small to justify continuous manufacturing. The demand for products may come from customers or other divisions of the company. The variety of work again requires skilled workers and general purpose machines, but the workers make enough parts to get better at what they do. The machines do not require frequent setups. Product Planning is challenged to schedule job lots that flow through the manufacturing processes with the most efficient use of resources. Quality Assurance may be able to assure quality with statistical sampling.

Continuous manufacturing uses the full capabilities of a company to produce a stream of products until the production technology becomes obsolete or the market loses interest. Special-purpose machines are designed for repetitive operation without much human attention. The use of automation is maximized, limited by the profit available to invest in complex machines. Product Planning must adjust schedules as machines fail or require preventive maintenance. Quality Assurance testing may be reduced by automated tests. Manufacturing Engineering is focused on reliability. Marketing is able to concentrate on applications of a range of qualities of the product for its customers' needs.

Twenty-five years later, the biggest change to Wright's version is in the way that Computer Aided Design (CAD) feeds directly to Computer Numerically Controlled (CNC) machines to reduce the need for skilled workers in custom and intermittent manufacturing.

Fluid Processes

The situation for fluid processes is very similar, except that custom manufacturing occurs in laboratories. Pilot plants make small quantities, but they are not custom. The pilot plant is a small version of a batch or continuous process that is used to discover the problems that occur when scaling up a process developed by R&D. Pilot plants have skilled operators using automatic control. Automation can be designed after it is known how to make the product outside of the laboratory, and fine-tuned by Manufacturing Engineering (or Process Engineering) as experience with the process accumulates.

Intermittent manufacturing becomes batch processing in the fluid process world. The number of possibilities for materials and setups is limited by the need to run pipes between process units. More flexibility can be had with pipeless batch processing, where a single vessel is made mobile and equipped with docking connections that fit fixed supplies of material and energy as well as analyzer stations. The batch vessel moves from station to station just as a discrete part moves from machine to machine. Many mobile vessels keep the static stations busy, as if the system was an assembly machine.

Continuous manufacturing or processing is subject to the same market fluctuations as discrete manufacturing, but there are some stable products, such as fasteners for discrete and gasoline for refining. Actually, the need for chlorine bleach may outlast oil, but the need for standard screws and bolts will be with us until civilization collapses.

Managing with Computers

The early use of computers led to the field of Computer Integrated Manufacturing (CIM). The use of computers to process business information depends primarily on the ability to reduce processing to algorithms. These can't be modeled without knowing the business organization. At the same time, the expense of modifying software applications tends to bring individual businesses closer to the organization that fits the application.

When IBM did its first development of applications that could be used to sell its hardware, it soon found that there was no such thing as a standard payroll accounting package, to give just one example. Customers were reluctant to engage in software projects, so they wanted to hear that the vendor had a standard package. IBM's business model was (probably, this is not written down anywhere) modified to sell the customer on the concept of a standard accounting package, and also to sell a service contract that would pay for a team of programmer analysts to create a custom accounting package that met the customer's needs. This was expensive, but nobody got fired for buying from IBM.

In another story from the life of a control engineer, a vendor who excelled at high level selling convinced a chemical company's executives that it had a system that could improve boiler combustion efficiency by 1.5%. This translates into serious money for large boilers. The engineering team that was designing a new plant was instructed to use this system. Attempts to reason with management failed. The engineers knew that you could get a one to two percent increase in the efficiency of old boilers, but not for new ones.

The new plant was built and the vendor's system was installed on the new boilers. The vendor had sold a service contract for the first year of operation based on their guarantee of savings. The boilers were first run without using the system to establish a baseline. The vendor's programmer spent a frantic year trying to meet the goal. There was some increase in efficiency, as there must be if you continuously measure residual oxygen in the flue gas, but not enough to meet the contract. Management, to avoid embarrassment by the engineer's predictions, declared the new system to have met its goal. If anyone learned anything at the management level, the engineers never found out. The vendor is no longer in business.

One other thing—the vendor offered upper management a money-back guarantee with return of the system if it didn't perform. Management at that level did not appear to appreciate that the costs of installation, training, and so on preclude the idea of ripping out an existing control system and replacing it with another, with the same attendant costs repeated.

Evolution of CIM

There are approximately three stages in the evolution of CIM. The first one follows.

John W. Bernard, a senior automation consultant, lists five primary functions and their major activities for a manufacturing business in his 1989 book "CIM in the Process Industries." They are similar to those of Wright: Planning instead of Marketing, Financial Control instead of Financial Affairs, Operations instead of Production, Technology instead of R&D, and Support instead of Industrial Relations. This is closer to what we have today, except that Engineering had not yet been replaced by Legal in 1989.

Planning

> Marketing, Product Planning, Business Planning

> Economics, Pricing, Advertising

Financial Control

> Accounting, Treasury, Finance

> Data Base Administration

Operations

> Sales, Order Entry, Production Planning, Material Acquisition

Manufacturing, Assembly and Finishing, Quality Assurance

Billing and Shipping, Customer Service

Technology

Research, Development, Engineering

Technical Resources

Support

Human Resources, Security, Training

Information Systems, Maintenance

Here is another arrangement, with Planning, Financial Control, and Technology unchanged. The different capabilities for business and manufacturing are formally separated.

Business Operations

Sales, Order Entry, Production Planning, Material Acquisition

Billing and Shipping, Customer Service

Business Information Systems

Manufacturing Operations

Manufacturing, Assembly and Finishing, Quality Assurance

Manufacturing maintenance, Inventory

Process Control Systems and Simulators, Operator training

Support

Human Resources, Security, Personnel Training

Information Systems maintenance

The reason for separating business and manufacturing systems is that they require different ways of looking at the problems that they face. Business involves transactions which vary in timing and importance. Business communication systems and applications are designed around transactions. Manufacturing requires current process measurements as the basis for what to do next, where "next" can imply the next few milliseconds. Transactions occur at human speeds, in seconds to minutes to hours, depending on the time required to reply to a query. Manufacturing control systems do not deal well with delays. Dead time (delay) makes a feedback control loop less able to maintain a measurement at its setpoint. For discrete control, delays slow down the production rate.

The problems faced by process control and business applications are also quite different. The difference is enough to make it improbable that a human brain could understand both. Separation is required, with common goals and different methods. The human brain can't handle both kinds of problems because it only weighs three pounds, about 1–3% of human body weight. Your ego may tell you that your brain is as big as a planet, but this does not hold up in practice. Yes, brain cells are microscopic, so that there are over 100,000,000,000 of them in those three pounds, but those cells do not translate into ten gigabytes of memory. What makes the brain complex is the thousands of associations that each neuron has with other neurons.

The Second Evolution of CIM

Professor Ted Williams of Purdue led work financed by a consortium of users and vendors to define a reference model for CIM during the 1980s. The result was published as "A Reference Model for Computer Integrated Manufacturing" in 1988. The ISA published it in 1989, and it became known as the PRM. The Purdue Enterprise Reference Architecture (PERA) was created from the PRM. A companion manual was published with the intent of revising it as more businesses tried the PRM, called the "PERA Master Planning Handbook." The latest version at the time of this writing is dated 2006. Both documents are available at Gary Rathwell's web site www.pera.net, but they are for personal use.

PERA and the PRM are still the definitive reference works for vendors and users engaged in using computers to achieve consistency in the activities of middle management and supervision, as well as process control systems. Computers and software applications have had 20 years to improve on them to take advantage of increased computing power. ISA-95 is based on them, but it is useful to go to the source to see why the standard says what it does.

The functions of a manufacturing facility within an enterprise are divided into a hierarchy with five levels. Level zero, the basement, had no computers in 1988, so it was used as the lowest level, which sent no commands lower. Microcomputers in devices have since muddied the waters, but the concept remains useful. Levels 1 and 2 contain basic and supervisory computer functions with corresponding human functions for controlling a process. Level 3 contains functions that tie processes together to form the production function at a facility. Level 4 contains those business management functions that can benefit from computers.

The model stops at level 4 because higher levels of management require levels of innovation that computers had not attained in 1988 and still have not reached. A fifth level was added in a later revision to include automatable management tasks for coordination of different facilities. The sixth level for upper management is implicit, but not stated in a CIM model because there are no computers that think like that. There is no place in a computer to inject testosterone.

Appendix I of the PERA document, 2001 version, has a section on data flows. Figure AI-38 shows the data flows for a generic manufacturing facility that is one part of a larger enterprise. Such functions as Marketing & Sales, R&D, Purchasing, and Accounting are done at the corporate level. Other external factors that communicate with the facility are customers, suppliers, and shippers. Local management is required to deal with these external factors.

There are nine functions within the facility. Order Processing, Production Scheduling, Product Inventory Control, and Product Shipping primarily satisfy customer orders. If there will not be enough inventory, scheduling negotiates with Production Control to manufacture products to fill the order, after they pass Quality Assurance. Production also requests material and energy from Material & Energy Control, which gets what it needs from Procurement. All functions talk to Product Cost Accounting.

Figure 5.3 is an approximation of figure AI-38 in PERA, so as not to copy it. The relative positions are close to PERA, but all of the 50 or so data paths are not shown. Some paths are shown without identifying their contents so that Figure 5.3 will show some of the complexity.

The following table names the circles for manufacturing activities and the squares for external activities.

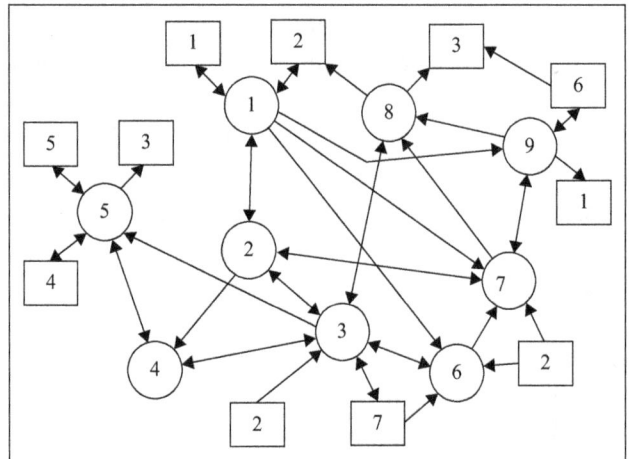

Figure 5.3. PERA data flows.

	Manufacturing		External
1	Order Processing	1	Customer
2	Production Scheduling	2	Marketing and Sales
3	Production Control	3	Accounting
4	Material & Energy Control	4	Purchasing
5	Procurement	5	Supplier
6	Quality Assurance	6	Shipper
7	Product Inventory Control	7	Corp R&D
8	Product Cost Accounting		
9	Product Shipping		

Each of the nine functions has one or two levels of sub-functions. For example, Production Control (3.0) has Operations Control (3.3) to actually make products, supported by Operations Planning (3.4), Process Support Engineering (3.1), and Maintenance (3.2). Maintenance, in turn, has Maintenance Crew Supervision as its primary sub-function, supported by Maintenance Planning, Documentation, Spare Parts, and, of course, Cost Control.

Figure 5.4 approximates figure AI-41 in PERA, again avoiding copyright issues. This figure shows the first layer of sub-functions for the production control function. The numbers in the smaller circles and squares are the same as for Figure 5.3 and are shown in the table. The two-way connection to level 2 carries commands or requests to the process control system and returns its responses.

This brief look at the functions in PERA shows the tip of an iceberg

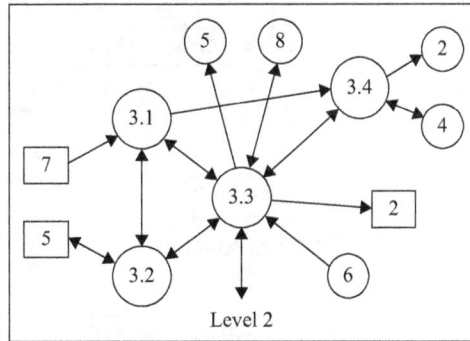

Figure 5.4. Production control data flows.

(a term that will be obsolete when there are no more icebergs). However, PERA and the PRM are about the use of people, equipment, and computers in manufacturing, which is outside the scope of this chapter.

The Third Evolution of CIM

The ISA-95 standard builds upon the PRM and PERA to specify models and methods for communication between manufacturing process control systems and business process control systems. The part of the standard that is of interest here is the section on Manufacturing Operations Management (MOM) models in Part 3 of ISA-95. The standard identifies activities and defines what they are to do, but does not define how to organize the tasks. However, you have to know what has to be done before you can organize the people.

Manufacturing Operations has four categories of activities condensed from activities in the PERA data path model. They are Production, Quality Assurance, Maintenance, and Inventory. All of them are described as being in level 3 of the PERA model, between the enterprise business activities in level 4 and the process control functions in levels 1 and 2. Level 3 contains the activities that convert business requirements into process automation commands or requests, and converts process information into business information.

Some organizations may put some of the level 3 activities into level 4 for business reasons (a term that covers many things) or into level 2 because those things are already done there. An activity should not move to level 4 if it affects product quality, is required for regulatory compliance, or is critical to plant safety, reliability, or efficiency.

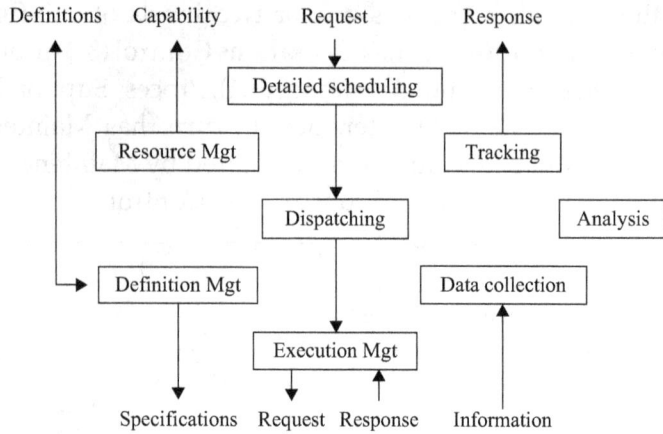

Figure 5.5. ISA-95 production model.

The ISA committee produced a common set of eight activities for all four of the MOM categories. The tasks within an activity change for each category but remain appropriate for the activity. See Figure 5.5. All of the activities may exchange data, so the only data paths shown are those to other levels and the main path from scheduling to dispatch to execution.

Activity begins with a request from level 4 to Detailed Scheduling. This activity determines the timing for acting on requests. Dispatching takes a request whose time has come and matches it to available resources and any required definitions or specifications. Execution Management converts the fully detailed business request to a series of requests or commands to level 2 or 1. Each request provokes a response from level 2 or 1 containing the success or failure of the request. Execution Management notifies Tracking and perhaps Analysis of each completed task.

Definition Management maintains a level 3 source of rules, procedures, and specifications required for business requests queued by Detailed Scheduling. It may request information from the corporate database in level 4, and it may provide information directly to level 2. Level 1 does not contain devices that can interpret specifications.

Resource Management keeps track of the human, material, and energy resources that are available, busy, or unavailable. It calculates a capability response for the category for use by level 4 activities and provides information to scheduling and dispatching activities. It gets information from Execution Management or Data Collection.

Data Collection receives status information from level 2 and puts it in a local database, perhaps with data compression, for use by other activities in level 3, not just for its category. Analysis processes collected data into forms suitable for level 4, such as key performance indicators, various resource utilizations, and information that will identify problem areas or suggest areas for improvement. Tracking processes data from other activities to form a business response to level 4, as well as keeping track of work done by the category.

ISA-95 contains models of best practices. If your organization is still searching for ways to organize the activities of manufacturing and business processes, this is a good place to start. The goal of ISA-95 is to standardize the communication between processes using variants of Extensible Markup Language (XML). Parts 1 and 3 are enough for looking at organizations.

Management Styles

Management styles have changed with the growth of technology. When education was rare, managers had to teach workers just enough to get their job done at a time when jobs weren't particularly technical. Workers, like all humans at the core, didn't want to work any harder than they had to, so managers had to use authority and threats for non-compliance to get as much work done as possible.

Technology workers tend to be driven by curiosity to work at solving problems. They work better with managers that they respect for their technical ability while being aware of their authority. In some cases, the authoritarian manager is one that doesn't have respectable technical competence.

Before computers, process operators had the mechanical competence to be able to diagnose equipment failures, but few understood the science behind what they were making. Their sense of the process came from gages and the physical senses of what the process was like when it was running well, and by experience of what it was like when something was wrong.

Computers eventually changed that by giving operators much more information about the process. Those that could use the additional information became quite good at operating. And why not, considering that they lived with the process as no one else did. The situation could be compared to a farmer and his crops, where the land is his process equipment and the weather his disturbances.

A Rosemount RS3 control system was installed in a research refinery of Conoco in Tulsa, OK, in the mid eighties. Operators were taught how to configure the system and its control loops. One operator reduced the energy required for a distillation column by 30%, by trying things during her shift. The engineers involved were surprised and allowed as how they hadn't thought of that. Of course, no engineer had sat with the process for eight hour shifts to gain the knowledge that comes from small experiments.

It is not possible to make even small experiments in the regulated industries. A great deal of work goes into validating the design, installation, and quality of a process. Operators are strictly there to handle abnormal situations. Their work to fix a problem will be carefully reviewed to see if the product was harmed or can be shipped. In other cases, authoritarian management has settled on the one true way to make product. An experiment could cause an operator to lose her or his job.

Managing with Smart Machines

The studies of computers and people by Zuboff in "The Age of the Smart Machine" were discussed in the chapter on operators. She also worked with managers in various industries. Her work with two paper mills owned by the same corporation is of interest here. Both mills installed their first computer control systems in the mid eighties and went through the adjustment from plant floor to central control room. The mills were identified in her book, but they are just A and B here.

The story begins with the plant manager of A asking his direct reports if they will all be working for a smart machine, or will they have smart people using the machine. Zuboff's interviews with managers revealed a desire to develop computer systems that would replace operators and provide a central point for management to give direction to the machine. She noticed that these descriptions ended with a ritualistic reference to people by saying that the machine will free up the operator's time for problem solving. There was rarely any mention of developing problem-solving skills in the operators.

Some managers admitted that justifying computer control systems to corporate management required a reduction in head count. One said, "We have simply looked at bodies rather than price per ton. We never asked the question, "Can I keep this person and get more tons?" Another said that they had just finished a large report on how the computers would talk to each other without ever mentioning how computers would talk to operators.

The difficulty seems to be the belief that workers do not need to know, they only need to obey. The alternative, that workers can know, questions the need for managerial authority. It also implies that workers need to know in order to do their jobs, "which is an affront to the collective faith that keeps a hierarchy afloat." Managers would become teachers, needing to persuade workers to work—giving up the easy authoritative unquestionable work direction—"Do it because I say so."

You see, it was not just the operators that were having trouble with automation.

Trust   One manager told Zuboff that they do not trust operators to work well with the computer, and they must be certain that the work will be done correctly. The operators were not blind to these feelings. One said that management is afraid that we'll sabotage it if we know how it works. Another said that they treat us like kids, showing no respect or trust.

Exactly! These are the kinds of things humans do if they have not developed trust. It seems as if this is impossible to do unless everyone works together to defeat a life-threatening situation. The central engineering management people at Hercules in 1960 were all war buddies who had learned to trust each other. The workplace offers few opportunities to establish trust at that level.

Harvey Mackay (Swim with the Sharks, 1988, and others) talks about "The Trust Edge" by David Horsager in the Star Tribune business section on Oct 8, 2012. Horsager says, "Without trust, organizations lose productivity, relationships, reputation, talent retention, customer loyalty, creativity, morale, revenue, and results." Mackay says trust isn't created overnight. Horsager says, "While it appears to be static, trust is more like a forest—a long time growing but easily burned down with a touch of carelessness." Mackay says a study of salespeople found that the top trait was honesty, not charisma, ability or knowledge.

Indeed, trust is a basic human capability. Children learn it before they are two, or find it very difficult to ever trust someone else. Trust is basic to reproduction and raising a child. You don't need trust if you have authority, but you won't get cooperation, just fear. That's why workers form unions. But you knew all that, just not in connection with automation.

Zuboff relates what happened when a process engineer and the manager of the paper-making group put together a computer application that would show the paper machine operators the current cost per unit of paper produced by the machine, called the Expense Tracking System (ETS). This was aided by direction from corporate management to develop "a more cooperative and flexible organization attuned to cost reductions."

Development of the ETS took a little over a year, with the help of operators who "contributed things no one else even conceived of." After implementation, operators gained increasing responsibility for costs and proper operation of the machine. They could see things that had previously been hidden from them. They could correlate behaviors of the paper machine with total cost of the product they were making.

Zuboff reports that the predicted savings for a year of operation were $370,000 and actual savings were $456,000. All this happened by poking a hole in the "knowledge dam." While numbers for the year looked very good, savings had reached a plateau at eight months, with no new insights. The process engineer knew that something was wrong, because there were more things that could be done, and that only a third of the possible savings had been reached.

There were several complicating factors. The operators, obeying the rule that no two people are alike, fell into three groups. About 20% of them caught on rapidly and made effective use of the ETS, 50% were not as quick but did become effective, and 30% seemed indifferent to the system. Some of the 30% could learn but didn't see a need for it, and the rest probably couldn't be trained. The process engineer was reassigned before he could start the next level of training.

Social Change   The real problem turned out to be the managers in contact with the operators. If the operator does something that loses money, the manager will be held accountable, but if the operator isn't free to experiment there will be no improvement. As one manager put it:

> Traditionally, managers bring discipline to their organizations around centerlines, documentation, sharing information. The computer means that [this] can be done without a manager. If we combine the computer with comments from the operators, most of these things can be accomplished. We are all struggling for our turf. The technology will mean change in our social systems, but so far this social change in management is not orchestrated.

The best operators agreed that the managers felt threatened, and tried to take credit for what the operators were doing. The operators didn't know how to make the managers feel better, so they stopped improving and went back to doing what they were told to do. They did that to save their jobs in a hierarchy that had the managers determining their pay.

One operator lamented that it was no longer possible to experiment with the paper machine, even on the night shift, but that is how you discover how to improve the system. Another said that the ETS had become a way for the managers to measure their performance, and if it is not good they look for someone to blame. One perspicacious operator said:

> All that managers care about is exposure. They want to do things so that other people up in the hierarchy.... will know that they are doing good work.... Once they found out that this expense tracking thing was going to be really looked at by corporate management, they started to latch onto that thing and they wouldn't let it go. They are looking for the glory from the system.

And so it goes with all hierarchies that attract and encourage those who climb the dominance ladder. See the book "Dinosaur Brains" by Albert J. Bernstein, or on the darker side, "Snakes in Suits" by Paul Babiak, or even the cartoon work of Scott Adams, creator of "Dilbert."

An operator relates that one of the higher-ups brought some customers into the control room for a tour. He showed them one of the new computers and said, "This is a marvelous machine. It is a wonderful piece of equipment. It doesn't take coffee breaks or come in with a hangover. And it does a better job than a man can do." All of this was said as if the men were not standing there, like they were not people. The manager was, as they say, unclear on the concept but the effect was not lost on the operators.

Zuboff says "As managers use computerization to escape their dependency on workers' unique skills, both managers and workers alike become more dependent on the computer system." The operators at one plant called their computer "Otto" (Otto Matic) as they humanized it to make it part of the group. Of course, Otto didn't always make sense as it followed its programmed applications, written by programmers who got their direction from people who didn't always understand the situation.

That was plant A, where experienced operators made the transition to computer monitors. Plant B was new, designed from the start to use young people not afraid of computers and train them in how to use the new technology. Computer applications included a "calculator model" and a Data Access System (DAS) that had a central database for all plant data, not just the process.

Teamwork was emphasized, so that no one acted alone without discussing it with the team, if the action would have a new effect on the process. New hires received training in basic math, chemistry, and physics as well as the process. Simulations of the process were available. Some people became designated sources for guidance and education.

Zuboff says that by the fourth year of operation it became apparent that "the plant leadership had underestimated the persistent features of organizational life that would inhibit a full exploitation of their technological investment." The same labor–management difficulties that plagued plant A appeared at plant B.

One problem was the calculator model. Rather than simply show the operator the unit cost of production, the calculator had a dozen different things it could calculate depending on what the operator chose to use as input data. This required the operators to know the models in order to know what influenced the results, but the models had "fudge factors" that were not known to them. One of the operators complained that the models couldn't be understood with what they had been given, saying, "As long as it's a black box to me, all I can do is babysit the computer."

Asked about the ideal manager, one of them said this:

In a traditional system, managers are drivers of people. You focus on driving people to work as hard as possible. With our new technology environment, managers should be drivers of learning. They should be driving everyone to use technology for better understanding and ways to expand the business.

Perhaps the problem was that the managers didn't know how to do that.

The preceding discussion of management styles was meant as a cautionary tale from events at the beginning of the computer control revolution. Some progress has been made. There is more talk of empowering the operator to make decisions. People are re-designing the operator–machine interface to improve it, but human nature is slow to change, and even the profit motive can only do so much.

Automation and Information

There are two ways to use a computer for process control. Automation allows the computer to sense and manipulate process variables directly. Information allows humans to supervise automation, providing that the automation is open to supervision and not enclosed in an impenetrable black box. We can automate processes and informate people, to use Zuboff's term.

Actually, process automation is quite open to supervision because the people who work with processes know that machines can't be taught to handle everything. This was not initially true in aviation automation. What the military did is unknown, but commercial cockpit automation began with Airbus and the joystick that replaced the yoke in a completely fly-by-wire aircraft. The aircraft attitude and engine controls were fully automated. As more experience was gained and some aircraft experienced controlled flight into terrain, it became clear that black-box automation was inadequate for the real world. Pilots found the aircraft doing things that they couldn't correct, or being limited in commands that they could give by something in the automation.

The joke going around at the time was that before long the only occupants of the cockpit would be a man and a dog. The man was there to feed the dog, and the dog was there to bite the man if he touched anything. This has been extended to the factory of the future. Indeed, the man had better not touch anything because lack of practice would make him incompetent to operate.

It is far better to keep humans in the loop, using automation to do the things that humans can not do, such as regulate hundreds of loops in a process or manually operate a milling machine to make complex shapes exactly alike for long periods of time. The operator uses information about the remote process to monitor its health and respond to problems when health degrades or fails, never losing touch with what the machines are doing, and always looking for ways to improve control of the process.

The Role of Ego in Management

James Collins wrote "Good to Great" in 2001, which became very popular. In it, he said that humility was one of the characteristics of great leaders. He also said that two thirds of the companies that didn't become great were held back by the "presence of gargantuan personal ego." Marcum and Smith wrote "Egonomics" in 2007 to expand on the degrees of ego that can turn a good leader into a corporate liability, and to talk about ways to find the right balance of ego and humility. They determined to write the book after discovering that "63% of business people say ego negatively impacts work performance on an hourly or daily basis, while an additional 31% say it happens weekly."

You may well ask, "What does ego have to do with automation, and why is it in the management chapter?" So far, automation schemes are planned and designed by humans. We all have to have enough ego to get out of bed and face another day, believing that there are important things to be done. The ego can grow if it is fed, and so people in leadership positions are at risk. The CEO surrounded by "yes" men is near the end of the line. At that point ego drowns any humility, blinds a person to reality, kills any curiosity about the changing business environment, and deafens a person to truth. The CEO refuses to accept any possibility of defeat.

These things can happen to a lesser degree to people further down in the hierarchy, who still retain some of their humility because there are more powerful people above them. They are vulnerable to spurts of ego, sort of like growing pains. We don't necessarily die when we lose control of our ego, but we can lose accumulated trust and respect. As discussed above in Managing with Smart Machines, loss of trust can be very damaging.

John Mordecai Gottman has done extensive research on marriage and family, which is equally relevant to business and employee relationships. He predicts the outcome of relationships by looking for four behaviors associated with excessive ego: stonewalling, defensiveness, criticism, and contempt. Basically, the ego has gone too far when changes in the circumstances that once favored a course of action cause the ego to deny that anything could be wrong. That is the time for some humility to allow others to discuss new courses of action without feeling attacked by the new ideas.

Country singer Mac Davis expressed the problem in 1980 in a song that begins, "Oh, Lord, it's hard to be humble when you're perfect in every way."

This is not the place to go any deeper into the subject. "Egonomics" covers the subject completely. The ego was discussed here because it is an important and overlooked attribute of management that can affect anything from an automation project to the future of the company.

Alternative Architectures

Zuboff has exposed the problems of combining computers and management hierarchies. Others have noticed that hierarchies are not working very well anymore. Dee Hock, who wrote "Birth of the Chaordic Age" in 2000, has said the following in different ways:

> Institutions are increasingly unable to achieve the purpose for which they were created, yet they continue to exist as they increasingly devour resources, demean people and destroy the environment: schools that can't teach, welfare systems in which few fare well, unhealthy healthcare systems, corporations that can't cooperate, economies that can't economize, police that can't enforce the law, judicial systems without justice, and governments that can't govern.

Not to mention "families far from familial," and "communities far from communal." He is well worth reading because he has ways to solve the problems. Another excellent author is Patrick Hoverstadt, who wrote "The Fractal Organization: Creating Sustainable Organizations with the Viable System Model" in 2009. Hoverstadt builds on the work of Stafford Beer, who used mathematical Operations Research to model organizations based on live brains, bodies, and environments. Beer wrote "The Brain of the Firm" in 1972 and "The Heart of Enterprise (Managerial Cybernetics of Organization)" in 1995. Large and small organizations have successfully applied their principles.

Jon Walker wrote a description of the Viable System Model (VSM) and his experiences in applying it in 1991 and published it on the web as "The Viable Systems Model." The latest version is V 3.0 dated 9 Aug 2006, and is listed in the references. The case histories are useful, but not discussed here.

The most striking feature of the model is a hierarchy of control that does not depend on or admit the competitive instincts of humans, with accompanying authority and demands for obedience. Since this denies reality, it hasn't exactly caught on, but it is instructive to examine it. The most important feature of the model is that it is built on a foundation of autonomous wealth-generating processes, not necessarily manufacturing. Once you have those, the problem becomes one of coordinating the processes so that they act for the good of the entire organization.

The foundation of autonomous processes requires intelligent workers, not beasts of burden that must be whipped into obedience. It recognizes that the people most closely involved with the process are those who know it best, and are the most capable of improving it. It also requires that the people involved have good information about their process, and are not making wild guesses at improvements. Tom Peters (In Search of Excellence, 2004) became famous writing about self-management and worker empowerment, and the need for it in an age of smart machines.

Consider that the autocratic hierarchy treats each worker as a problem that must be solved by giving orders and measuring detailed performance. This is necessary if the workers have a minimum understanding of what they are doing. If the workers can have enough information and training to solve their own problems, and the autonomy to do that with minimal supervision, then one supervisor can monitor many workers. The VSM requires autonomous workers.

The VSM contains five systems and their environment, as shown in Figure 5.6. System 1 contains the Operations that make the money required to keep the organization alive. Each operation functions autonomously, as long as it can handle its own problems. The other four systems act to regulate system 1 so that the organization stays in business. Beer calls this collection the Metasystem. Following the body metaphor, system 1 is the body, the metasystem is the head, and the diagonal lines to the right are the spinal cord.

System 2 is required to monitor the operations in system 1 for abnormal behavior, so that no operation is allowed to fail without giving warning to the business. If possible, system 2 resolves the problem in a way that is acceptable to the people in the failing operation. Complex operations involving more than 6–12 people may have their own small–scale metasystem that condenses data for presentation to system 2.

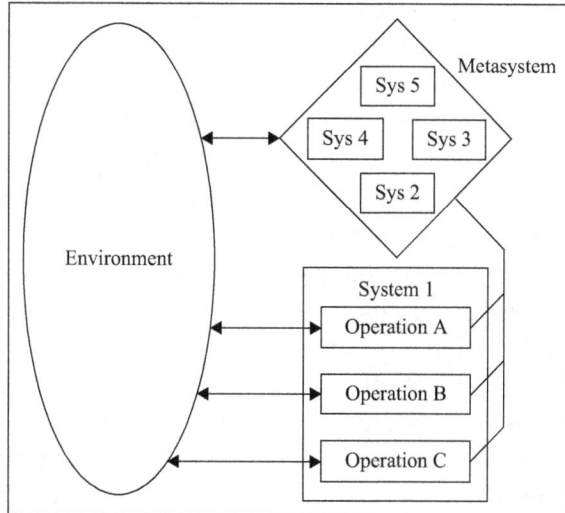

Figure 5.6. Viable system model.

System 3 resolves problems that system 2 cannot handle, looks for ways to optimize system 1, and discovers opportunities for synergy among operations. System 4 looks to the future by sensing trends in the organization's environment and makes plans for the future based on the organization's capabilities. System 5 provides an ultimate authority, a place where passing the buck or kicking the can stops. It provides ground rules, policy, and corporate identity, but these are approved by people in the other systems.

The components of system 1 are not completely autonomous. They have agreed to a plan of action, and are measured by the activities of system 2. They have also agreed on the conditions that will cause them to lose their autonomy and become subject to decisions from system 3. System 1 is not subject to arbitrary decisions from above unless the organization as a whole is threatened.

All of the systems must be in balance for the organization to be viable. Balance means that there are as many ways of sensing and controlling variables as there are variables. Beer calls the number of variables "variety." Balance is equivalent to having a process control system for as many variables as the process designer decides are necessary. This means that system 4 must monitor as many environmental variables as might affect the organization. System 3 must have a complete model of the organization in order to optimize it, and so on.

System 5 is where the difference between authoritative and cooperative behavior comes to a conclusion, and where the whole thing goes against human dominance hierarchy behavior. If the people involved can get over their stone-age nature, a marvelous structure emerges.

The systems of the human body cannot compete; they must cooperate. If a competing system won, the rest of the body would die. Competition for dominance evolved as humans competed for territory as their populations grew. This wasn't new with humans, many animals including chickens compete for dominance, but the population of lower animals is controlled by predators. Uncontrolled population growth leads to extinction as resources are consumed, so natural selection favored those organizations that could kill off enough people to relieve the pressure on the resources. Technology has blunted the effect of war for that purpose, while it accelerated population growth. Competition is obsolete, but still built into human nature. Might as well imagine world peace as humans without competition.

The VSM is recursive, in that the operations in system 1 can be manufacturing processes, plants containing those processes, regions containing plants, or a multinational organization containing regions. The model works with as few as five people, providing each of them has the talent to do one of the functions of the five systems when required. Normally all five do what needs to be done, to produce income, but all five can meet to solve a production problem or set policy. Recursion allows expansion to multinational organizations or nations.

The major problem, as Zuboff has shown, is the inability of management to grant autonomy to parts of system 1. Walker says, "The management gurus in the USA are now (1991) coming out with statements like, 'Train the hell out of your workers and then get out of their way.'

Walker gives an example of what goes wrong when there is no autonomy. The city has torn up his driveway as part of another project, and is now replacing it. Workers have installed wooden forms outlining the driveway. When the truck with the blacktop (tarmac) arrives, it is too wide and smashes the wooden forms. The blacktop is laid anyway, because that's what the workers were told to do. Doing anything on their own is totally alien to their culture. One of them said, "We're all supposed to be totally stupid and only able to do what we're told." That kind of obedience requires someone in authority to monitor the operation, but that person wasn't available.

The VSM is presented here to encourage alternative thinking. It is much better adapted to the age of the smart machine than the hierarchies of the last three centuries. Still, there is the nagging suspicion that not enough people will be capable of autonomous action to make this work. There is a reason why hierarchies evolved. Automation may make it possible, but then there is the problem of what to do with all of the people without jobs.

References

A Reference Model for Computer Integrated Manufacturing, 1988, www.pera.net

ANSI/ISA-95.00.03-2005-Enterprise-Control System Integration, Part 3: Models of Manufacturing Operations Management, 2005, www.isa.org

PERA Master Planning Handbook, 2006, www.pera.net

Babiak, Paul. *Snakes in Suits*. Harper Business, 2006.

Beer, Stafford. *The Brain of the Firm*. Wiley, 1972.

Beer, Stafford. *The Heart of Enterprise*. Wiley, 1995.

Bernard , John W. *CIM in the Process Industries*. ISA, 1989.

Bernstein, Albert J. *Dinosaur Brains*. Wiley, 1989.

Collins, James. *Good to Great*. Harper Business, 2001.

Hock, Dee. *Birth of the Chaordic Age*. Berrett–Koehler, 2000.

Horsager, David. *The Trust Edge*. Summerside Press, 2011.

Hoverstadt, Patrick. *The Fractal Organization*. Wiley, 2009.

Mackay, Harvey B. *Swim with the Sharks without Being Eaten Alive*. MJF Books, 2005.

Marcum, David and Smith, Steven. *Ergonomics* Fireside/Simon and Schuster, 2007.

Peters, Thomas. *In Search of Excellence*. Harper Business, 1982.

Walker, Jon. The Viable Systems Model, 2006. http://www.esrad.org.uk/resources/vsmg_3/screen.php?page=home

Wright, R. Thomas. *Processes of Manufacturing*. The Goodheart-Willcox Company, Inc., 1990.

Zuboff, Shoshana. *In the age of the Smart Machine: The Future of Work and Power*. Basic Books ISBN 0-465-03212-5, 1988.

Automation

The first rule of any technology used in a business is that automation applied to an efficient operation will magnify the efficiency. The second is that automation applied to an inefficient operation will magnify the inefficiency.

—Bill Gates (1955-)

Introduction

There are three kinds of automation: manufacturing, office, and service. Office automation computes numbers and prints reports such as bank statements and progress reports. Service automation handles transactions, answers phones, and operates automated teller machines and airline check-in kiosks. There are other kinds of automation, but this book is about manufacturing.

Manufacturing automation is primarily used to overcome the limitations of human workers, increasing productivity and quality while reducing scrap or waste. There are situations where humans are at risk from the process, causing the cost of insurance to exceed the cost of automation. Mechanization, if not automation, is required when humans cannot lift or grasp objects, or handle them fast enough. Automation, in turn, is limited to the things that can be done with computers, but these limits are constantly being overcome as better hardware and software are developed. It is said that commercial aircraft flight control could be fully automated, but people prefer to fly with a pilot who cares about achieving safe landings.

Manufacturing automation controls the motions necessary to make products and intermediates. Factories do work over some period of time; work is defined as force times distance; distance over time implies motion. Discrete processes use motion control to automate the motions of machinists using Computer Numerical Control and Programmable Motion Control, both programmable positioning systems. Fluid processes use pipes, valves, and agitators to automate the motions of men carrying buckets and paddles. Pipes are passive, but they control the location of the fluid in them. The motions in discrete processes are generally visible, while the motions in fluid processes are not. The motions must be controlled to be useful. Discrete motions are controlled by mechanisms and motors. Fluid motions are controlled by pumps or gravity and valves.

This chapter will discuss the various kinds of motions and ways of controlling them so that processes may be automated. There will be some discussion of developments in other fields. The possibilities for autonomous control of processes will be investigated.

Definition

There are too many definitions of automation. In particular, those that tie automation to computers have oversimplified the situation. Webster's Tenth Edition (1996) defines "automation" as a noun (ca. 1948) derived from "automatic" with three usages:

1. The technique of making an apparatus, a process, or a system operate automatically.

2. The state of being operated automatically.

3. Automatically controlled operation of an apparatus, process, or system by mechanical or electronic devices that take the place of human organs of observation, effort, and decision.

The definition on the web today at http://www.merriam-webster.com/dictionary/automation is the same, except that in usage 3 the words "organs of observation, effort, and decision" have been replaced by "labor."

Webster's definition of automatic is unsatisfactory, but leans towards "self-acting." The definition on the web for English learners is more familiar: "... having controls that allow something to work or happen without being directly controlled by a person."

Automation generally means making something move by itself. Sometimes a computer is required to make mechanization be automation, sometimes a mechanical clock is automation, and sometimes it has to be something that a human couldn't do—for lack of strength, patience, or speed of response. You have one or more computers in your car if it was made in the last twenty-five years, but the car doesn't do anything that you don't command it to do with hands or feet. Is the car automated? Probably not—the engine, brakes, and steering are all controlled, but they don't act autonomously, at least not yet. A car that requires no human input to get from one place to another is automated, and has been proven to work, but people aren't ready for it.

When used as a noun, automation means a set of technologies used to animate equipment. When used as an adjective, as in automation professional, it means one who applies those technologies. According to the ISA, automation professionals are responsible for the direction, definition, design, development, application, deployment, documentation, and support of systems, software, and equipment used in control systems, manufacturing information systems, systems integration, and operational consulting.

Types of Automation

There are three types of automation: fixed, programmable, and flexible.

Fixed automation is expensive to change. A process or machine is configured to make one product, possibly with minor variations. The product needs high demand to justify the cost of automation. Examples include continuous fluid processes, parts of automotive production, and some packaging machines.

Programmable automation is time-consuming to change, but does not require new equipment. The cost is justified by the amount of product sold before re-programming is required, such as an automobile model year. A fluid process may have a piping system with valves that allow transfer from a selected source to a selected destination. Several transfers may be in progress if there are no conflicts in valve positions. An assembly machine or line may use the same equipment to assemble different products.

Flexible automation is not time-consuming because the programming work has been done in advance. The equipment follows different procedures to make different products, which requires flexibility as well as programmability. The cost is justified by the variety of products that can be made to meet changing demand, as is the case with some pharmaceuticals. These are the characteristics of a fluid process batch facility or a machine shop with CNC machines and robot arms.

Automation Technologies

The evolution of automation has five stages. Four of them have already started. Three of them are mature. First, there had to be machines, because one does not automate people. Second, there had to be controls for the machines, to reduce the variance in products made by the machines. This required the development of sensors and actuators as well as controllers. Third, there had to be digital computers to digitize sensor data and apply it to control algorithms, some of which could not be solved with analog devices. None of the stages are complete, as technology continues to drive change.

Stage One—Machines and Power Sources

A machine is something that directs or transforms power to perform a specific function. The lever is a simple machine that amplifies power in order to move something too heavy to move otherwise. An electric table saw transforms electric power to rotation of a saw blade for the purpose of cutting something that the saw blade can cut. The steam engine led to other kinds of heat engines, as pistons led to turbines. An industrial laser has no moving parts but can cut materials that a laser can melt. A machine may consist of many simpler machines that act together to perform a function. The technologies required to build machines with moving parts are a hundred years old or more, with continuing improvements. Lasers and computers are examples of more recent technologies for building machines.

In manufacturing, machines do things that are required to make finished products. Elsewhere, machines transport people and things, plant and harvest crops, generate electricity, and so on. Each machine has evolved from its first form to become better at what it does, depending on your definition of better. Commercial aircraft transport more people than ever before because the companies treat their customers like cattle. Technology changed rail transport from steam to diesel-electric. A less costly power source will change that again. There are machine tools that cut or weld with lasers in intricate patterns. The parallels to the evolution of life are direct, with new species arising from older species, and extinction for others.

Power Sources Machines require power to produce motion. Mechanical devices convert natural power (wind and water) to rotary motion, but are limited in the amount of power that can be produced in a device of reasonable size. Natural fuels can be converted to heat to produce steam to produce rotary motion, which can directly drive machines or generate electricity. Electrical power systems scale up nicely as they can have many sources in parallel through a distribution system.

Natural fuels were limited to wood until coal and oil were extracted in commercial quantities. Oil was used to fuel agricultural machines, which created surplus food. Parkinson's Law of population says that the population will expand to fit the available food supply, and it did. The population is now consuming as much of the natural fuel as it can, which has led to talk (but not much action) about sustainability.

Stage Two—Controls

Control technologies developed more slowly than machine technologies. Control as an engineering discipline developed from the use of servomechanisms in machines as the mathematics for stable feedback control were refined. Optimal control theory appeared around 1950, and robust adaptive control theories followed twenty years later. Robotics and the inverted pendulum further developed control technologies to the point where some schools taught control as a separate discipline from mechanical and electrical engineering. Increasing computing power per unit cost led to practical applications of the theories that further strengthened control technology as a unified field.

Devices required for control have evolved from simple proportional controllers with non-standard ways of sensing the values to be controlled and ways of controlling them. Their environment changed with analog and then digital signal standards and they evolved to suit their new environment, first with pneumatic signals and then with electric and now with wireless. Control equations once solved with marvelous force-balance mechanisms yielded to electronic amplifiers. Sensors made the change from pneumatic to electronic, which allowed more freedom to sense things that could not be sensed with air. Control valves continue to be pneumatic as the lowest cost reliable solution, but variable frequency drives for pump motors are gaining with lower energy costs.

Figure 6.1 compares traditional valve flow control with variable frequency drive flow control. The diagram with a valve has a conventional pump, motor, and motor starter connected to the fixed frequency power mains. Flow is controlled by throttling the valve, which produces a pressure drop, which

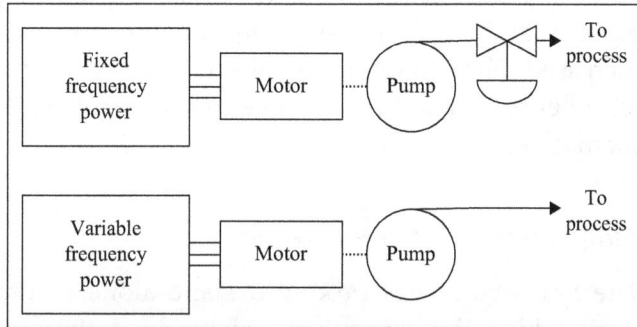

Figure 6.1. Efficient flow control.

wastes energy as heat lost to the environment. The diagram without a valve controls flow by varying the speed of the pump with a variable frequency drive. No additional energy is lost, except in converting power from the mains to variable frequency power, which is highly efficient and getting better as semiconductor switches improve.

Stage Three—Computers

Computer science began with the development of automatic calculators by Alan Turing and John von Neumann in the late 1930s, followed by a series of one-off designs that settled on stored program machines. At that time, people who used calculators to solve complex sets of equations were known as computers. The digital computer is clearly an example of automation, because it replaced human computers that could not compete for speed and accuracy. Various digital computers went into production after 1950 as computer science became computer engineering.

Introducing the computer into the control environment was sort of like introducing oxygen to the early atmosphere of Earth. There was an explosion of new control species as digitized sensor data could be processed in many new ways. The devices that were made before computers could not be completely compensated for changes in pressure and temperature, but compensation could be computed with ease. This especially benefited analytical devices.

Sequential and procedural control became much easier because the computer was designed to follow procedures. Process alarms could be managed with algorithms. The human-machine interface could be tailored to fit the human to the process. HMI design is also evolving towards standard methods, although humans introduce a wide preference for styles into the designs. Safety systems can accommodate a wider variety of risks, and also have standards.

Perhaps the greatest contribution of the computer to process control has been digital communication among computers for the exchange of sensory and other data in real time. This was first used for field devices in the early eighties and grew as the computations per watt ratio of microprocessors increased. Now many kinds of sensors can communicate digitally. Some use left-over processor time to solve control algorithms and manage asset data for themselves.

Stage Four—Communication

The first two stages produced stand-alone automated machines. The third stage added the computer, which added the possibility of communication among machines. Soon there was talk of connecting islands of automation to achieve controlled processes, and that has been accomplished with the aid of communication standards.

The fourth stage connects the manufacturing computers to the business computers so that business people have the data required to effectively direct production and the means to do so. Standards are again important, especially ISA-95.

Digital communication made it possible to put microprocessors in field devices so that sensor conversion and simple control could be done in field devices. A single standard for field communication is not possible at this early stage in the technology, because there is not enough user demand to sort out the competing methods. There may never be a single standard for discrete and continuous process control communication because the timing requirements are separated by a factor of a hundred or so.

Stage Five—Autonomous Operation

The fifth stage isn't here yet, for it requires computers that can create new solutions to new problems. There is uncertainty about the ability of a binary adding machine to achieve consciousness. Today's computers are constrained by the fact that they can only do things that they have been programmed to do, although some of those things are remarkably life-like.

Situation Awareness

As control became automation in the eighties, there arose problems with automation leading to fatalities. The most public problems were due to cockpit automation in commercial aircraft, where glass screens replaced individual gauges in addition to autopilots and autothrottles. Cognitive scientists deduced that the problems being called pilot error were due to a loss of Situation Awareness (SA) due to cockpit automation, also known as fly-by-wire.

The pilot of a manual aircraft uses all senses to be aware of what the plane is doing now and what it will probably do next, using a mental model built up by training and experience. Fly-by-wire eliminated the feel of the aircraft control surfaces through the stick. Full automation required mostly visual senses and interpretation of data presented on displays. If this sounds familiar, it is similar to what happened to Zuboff's paper mill operators in Chapter 4 "Operators."

In "Automation and Situation Awareness" (1996) M. R. Endsley writes, "Contrary to the implication of the term 'automated,' humans have remained a critical part of most automated systems." This is because humans have to monitor the automation for failures and conditions that it was not designed to handle. In some cases, humans have to perform some tasks that were not automated as well as monitor automation. Endsley also says, "Even though full automation of a task may be technically possible, it may not be desirable if the performance of the joint human-machine system is to be optimized." This statement neatly sums up the educational purpose of this book, and so you will find it in the front of the book.

D. A. Norman, in "The 'problem' with automation . . ." (1990) says, "The problem, I suggest, is that the automation is at an intermediate level of intelligence, powerful enough to take over control that used to be done by people, but not powerful enough to handle all abnormalities. Moreover, its level of intelligence is insufficient to provide the continual, appropriate feedback that occurs naturally among human operators." He suggests that the automation should be made either more or less intelligent because "the current level is quite inappropriate."

In case you are thinking that a process alarm is all the operator needs, Norman says that we do not know how to give appropriate feedback. "We do have a good example of how not to inform people of possible difficulties: overuse of alarms." Norman does not say that an alarm is a surprise, possibly one that destroys the mental model of what will happen next but does nothing to help build the correct model, especially if the alarm is not alone. ISA-18 goes a long way to reducing alarm overload, but they are still alarms and not a continuous modification of the operator's mental model of the process. Think of the SA difficulties of a train engineer going around a bend and seeing that there is no bridge where a bridge ought to be.

At the conclusion of Norman's paper, he says that there is a "perfectly natural" reason that adequate feedback isn't given to humans. There's no human feedback because the automation system itself doesn't need it. The design task is completed when the desired automation capability is provided. The designer sees no need for human feedback. Norman says that appropriate design should assume the existence of error and be able to "interact" with operators even in the worst-case situations. "What is needed is a soft, compliant technology, not a rigid, formal one."

Don Norman and Mica Endsley are still publishing papers about the same problem. An aircraft recently crashed because the pilot trusted the autothrottle to maintain approach speed, but it may not have been engaged. There was nothing in the cockpit automation to warn the pilot that the landing would not be successful. Perhaps there wasn't enough money in the budget to add the warning to the automation. In any event, SA is still a problem.

Details of the Five Stages

The following provides more detail on the stages of automation, for those that are interested.

Machines

Humans can move in many directions, which gives them great flexibility when manufacturing parts and assembling and packaging products. Machines (excluding robots) have just enough flexibility to get the job done, either automatically or when used by a human. A machine is composed of simple machines and mechanisms. The mechanisms are designed to provide the necessary motions, using gears, cams, linkages, and so on.

Manufacturing parts that are too heavy for humans to lift, such as heavy construction equipment and large engines, would be impossible without cranes like the giant overhead cranes in steel mills. These cranes need smaller cranes to assemble them and larger cranes to lift them into place. They are simple machines, but we would be severely limited in what we can do without them.

We think of gears as round circles that have fixed rotation ratios, but it is possible to use carefully designed shapes and placement of teeth to have variable ratios during one rotation, or even to have the output gear change direction. See http://www.mekanizmalar.com/ Cams are various shapes that are rotated by some power source. A cam follower translates the motion of the cam surface into the desired motion. Many cams may share the same line shaft to provide coordinated motions. See http://en.wikipedia.org/wiki/Cam. Manufacturing odd gear shapes and cams is expensive, so the combination of a stepping motor and a programmable motion controller has mostly replaced cams and odd gears.

Linkages are made of rigid bars (the links) fastened at their ends or elsewhere to allow rotation or sliding motion. At least one of the links is fastened to the surface that provides the reference for the resulting motion. This link may be driven by a stepping motor, while links attached to it provide the desired motion. It is possible to get straight-line reciprocating motion from rotation. Four bar linkages are very useful. See http://en.wikipedia.org/wiki/Four_bar_linkage. The cylinder, piston, rod, and crankshaft of an engine form a four-bar linkage. The pantograph is also widely used. There are large catalogs of four-bar linkages organized by the kind of motion they provide, as well as animations on web pages.

German engineer Fritz Reuleaux built 220 models illustrating mechanisms and their motions in the late nineteenth century. These were acquired by Cornell University in 1882, and assigned to 25 categories. Machine builders use catalogs like this to select mechanisms. See http://kmoddl.library.cornell.edu/model.php?m=reuleaux.

Discrete processes and packaging machines are constantly in motion while operating. Most fluid processes have only three moving mechanical parts—pump rotors, agitators, and valve stems. All of the other motions are made by fluids as directed by pipes and vessels, until the fluid is packaged for transportation or use. A fluid process is boring to watch compared to most discrete processes.

Control

There are two layers of control, basic and supervisory. See Figure 6.2, where the lower layer consists of basic control, sensors, actuators, and the process. The lower layer is designed to maintain a physical state or position in the process in the presence of disturbances. If the behavior of all of the disturbances can be predicted mathematically, then open loop control is possible. Usually the disturbances can't be predicted and have to be sensed. The information from the sensors is fed back to the controller, which operates on the difference between the desired state and the sensed state to correct the sensed state. This feedback control system is always reacting to an error and so is not able to maintain an

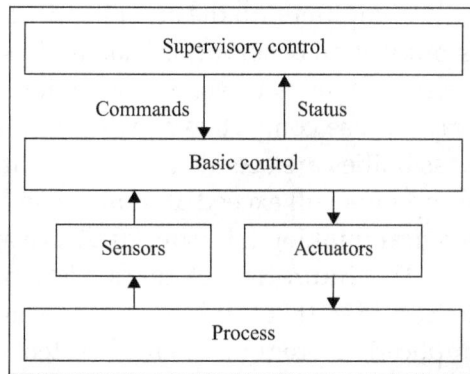

Figure 6.2. Supervisory over basic control.

absolutely steady state, but it does well enough for all practical purposes. In some cases, the behavior of a disturbance can be predicted and that information can be fed forward to the controller for better control.

Lower layer control of fluid processes can usually be done by general-purpose devices such as PID controllers because what they do does not vary among different processes. Some process variable sensors have long delay times before the information becomes available to a controller. Disturbances have shorter time constants and require a second layer of control (or deeper) in a cascade configuration. Control may be analog or digital, but modern controllers are digital because they can communicate more information to the upper layer.

Discrete processes use feedback control, but the feedback may be from the state of a switch, and may cause the process to go to a different state. Lower layer controllers are required to solve Boolean equations. This can be done with relays and timers, but the advantage of communication drives present solutions to computers such as PLCs. Digital position feedback encoders require digital controllers at this layer.

The upper layer, sometimes called supervisory control, adjusts setpoints in the lower layer controllers. This is done because the process may require states to change in a directed fashion according to a procedure that may be affected by what actually happens in the process. Upper layer control requires a general-purpose programmable device, such as a computer.

Computers

Control alone is sufficient to maintain the state of a process, but the process is either static or repeats a series of states in a simple pattern with few branches. A controlled process is simple enough that some humans can understand it and operate it. An automobile is a controlled process. The limited behaviors of control devices and controlled equipment have been relatively stable for decades and are not likely to undergo rapid change. Their possibilities are largely defined.

Computers, on the other hand, are very much in flux. Moore's Law (which is more of an observation than a physical law) has computer power doubling every two years or so. Software developers are barely able to keep up. New ways to use computers start new businesses with appalling frequency. Their possibilities are seemingly endless, and there are futurists willing to say that computers will exceed all humans in intelligence. If they exceed us in ways of magical thinking (Hutson, 2012), we are in deep trouble.

The future is not the subject of discussion at this point. It is useful to consider that computer hardware is not as important as the software application programs being created for it, and that the subject of this book is automation. Here, computer applications have been developed to use the underlying controlled equipment layer to improve manufacturing processes.

Some fluid process control applications can develop models of processes across many process variables, such as Dr. Charles Cutler's Dynamic Matrix Control, that reduce or eliminate the effect that controlling one variable has on another. Other applications can find the optimum operating point for a continuous process. Statistical process control can analyze patterns in batch processes to optimize key batch parameters. These applications began more than twenty years ago and have improved as computer hardware has improved. New applications are being developed, sometimes shortening the twenty years traditionally needed before theory becomes applied in an operating process.

Discrete processes have undergone amazing changes, such that a machine designer can use a computer terminal to design a device for production and then test the design with simulation applications. When the design appears to be ready for production, applications can create the drawings and bills of material for the device, and can program tools that make and test the parts for the device. Assembly may also be automated by programming the motions of robotic tools. This is too expensive for the small machine shop, but is being used for automobile and aircraft production.

Challenges Created by Computers The problem with computers is that users who are not programmers must interact with application programs created by programmers who are translating requirements given to them by marketing, who got them, they think, from the real users. The fact that this doesn't always work to everybody's satisfaction is attested to by the number of discussion lists that try to help users solve problems with applications and their documentation. The use of on-line help is diminished by the fact that the help text is focused on a fragment of the application, and assumes that the user knows what else is going on that might affect the fragment. For example, a help dialog for adding a process variable to a trend display assumes that the user knows how to create the trend display. A help dialog for creating a trend display assumes that the user knows how to access the configuration database, and so on. The answers are technically correct but useless without more of the fragmented information. Help dialogs are intended to be used as reminders of information already learned. A book that starts a topic at a simple level and increases in difficulty is required, but design teams are not charged with writing such books.

Further complicating the situation is the fact that vendors can't create custom applications to suit what a user has been doing to accomplish the task automated by the application. Single programs must accommodate many users with diverse requirements. Soon, one person's bug is another person's feature. The program grows in size and complexity as marketing adds features and benefits for prospective customers until it becomes untestable due to the number of combinations of options. Worse, the original team of programmers that wrote the program went on to do more interesting things and maintenance programmers, who don't fully understand why the designers did what they did, start introducing bugs by making fixes to the original code. At some point, the program becomes unusable and a new team is assembled to build a new version from scratch. New documentation is issued and the cycle begins again.

Worse, it is possible to make an application so complex that the user doesn't understand what the computer is doing, and has to experiment to determine the application's behavior. This is called opaque automation. The problem becomes acute in manufacturing automation when exception applications do unexpected things. Normal process operation is well understood, but when a combination of alarms occurs that has not happened before and exception handling appears to be making the situation more dangerous, the operator will lose precious time trying to understand what is happening. Opaque automation has killed people in aircraft.

All of this would go away if users could describe their applications directly to the computer. Programmers would be necessary to write computer programs that can interpret the user's needs with "creation modules" that assemble the required application. This leaves the user responsible for training others in its use, as well as continuous improvement and maintenance, but there is no one better suited to the task or more interested in getting it right. This will be far more effective than several layers of translation, in the same way that "what you see is what you get" programs simplified graphic design. The thinking required to create such modules has not advanced to practice, perhaps because programmers are quite happy with being necessary, but more likely because management is quite content to continue doing things the way they've always done them.

Data Storage Challenges Computers can generate lots of digital data from process sensors. Some people feel that they must save all of that raw data because they can. Raw process data has a time value of at most a few days, for tracing the cause of an alarm or upset. Raw data may be used to produce "signatures" of proper operation, again for diagnostic purposes. Once the signature is captured, or the process runs a day without an upset, the raw data has no value. The best way to capture upsets is in a rotating buffer, where newest data replaces the oldest. After an upset has occurred and things have stabilized, the whole buffer may be saved for analysis at a later time.

The sheer volume of data prevents it from being useful to humans. Instead we use statistics and data compression algorithms to reduce incomprehensible quantities of data to numbers that can be examined for patterns. Batch and discrete processes must log events during processing. Batch processes condense the events into a batch report that captures everything that might affect the quality of the product and that may be discovered at a later date. The report has no value after the product reaches the end of its life.

The computer in a PLC processes all of the Boolean equations from start to end fifty or more times per second. Sometimes it is necessary to save a previous contact state, but no attempt is made to capture all of the sensor data and save it. That isn't necessary because the equations are solved for the present state of the process. If there is a computer in the layer above the PLC that interacts with it, that computer may make predictions based on some of the past behavior. Statistics may be used to predict the life of equipment, and production reports may be prepared.

Think of it this way—each sample of the process conditions is immediately useful to calculate necessary changes to the process. Some of that data may be added to running averages. Then the next cycle occurs, and everything is done again as the unsaved and unnecessary data from the previous cycle disappears from the present. If the data from each cycle were saved for processing at some future time, it would take time to do that processing. When it was done, you'd have the same statistics that have already been gathered and the same reports, but no product would have been made by that processing. What was the point of saving all that data? Besides, you don't want the bandwidth penalty of sending a fire-hose stream of raw data to an archive. See the comments on Big Data in Chapter 10 "The Future."

Communication

There are usually four levels of communication in automated systems. The first level is basic control, which carries messages between process-connected sensors and actuators, and the computers that control them. The second level is operations or supervisory control, which carries commands and responses among the controllers and supervisory or human interface computers. The third level is manufacturing information systems, which contains computers and applications that aid management of manufacturing operations. The fourth level contains the business systems required to manage the business. See Figure 6.3.

The levels each contain their own networks for communication and a means for communicating with the next level up. There is some conflict when the corporate Information Technology department asserts that only IT has the knowledge required to maintain computers and networks. This attitude began when computers were esoteric devices known to few people. Now there are teenagers who can build computers and assemble networks. Some of them relieve their teenage angst by breaking into corporate computer systems just to see how far they can get. It is much more important to separate business and manufacturing systems based on the skills required to use them and the kinds of problems addressed rather than the hardware used.

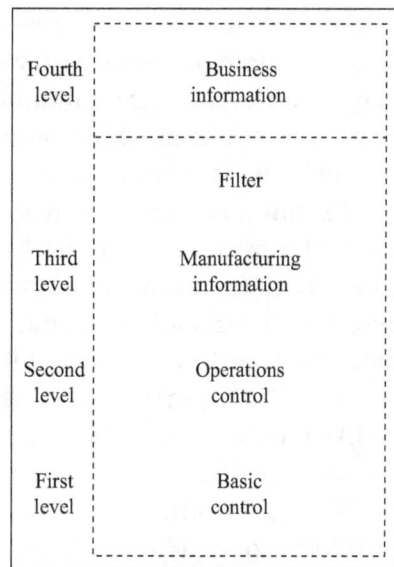

Fourth level	Business information
	Filter
Third level	Manufacturing information
Second level	Operations control
First level	Basic control

Figure 6.3. Communication levels.

It is necessary to have some sort of filter between the third and fourth level, or otherwise assure that no one in the board room can succeed in an attempt to change the settings of a control system. There was a time when system marketing departments pandered to the kind of ego found in a board room to promise that they could operate devices on the factory floor from high in their tower. If the manufacturing executives are not trained in operating the process, then a filter is required between levels three and two.

This concern with allowing only trained people to operate a process goes back to the days before computers, when a yellow line on the floor six feet (two meters) back from the control panel was as close as untrained people could get to the process controls.

First Level The first level carries data from sensors to the controllers and computers in the lower control layer, and carries data from controllers back to process-connected actuators. The signals were currents or voltages until microprocessors were used in the sensors and actuators commonly used for fluid processes, beginning in the late eighties. Discrete switches and solenoids communicated directly with PLCs until remote I/O using RS-232 and 485 serial lines grew popular in the seventies. Now we have wireless networks for some sensors.

Discrete and fluid process data are very different in their size and transmission rate. Discrete data may be transmitted as a single bit of binary data for two-state sensors and actuators like switches and solenoids. Position data is more complex, but both types have to be updated between fifty and a thousand times per second depending on the reaction time of the process. The rate will increase as machines become faster. Fluid process reaction times are limited by sensor conversion rates and the time constants of pneumatic valves, which may be measured in seconds for the large valves found in refineries. Data is seldom refreshed faster than ten times a second, and temperature data may be refreshed once a minute. This is a good thing, because there may be thousands of bits in a fluid process message.

In both cases, the data becomes stale at the refresh rate. There is no time to retry a communication if a message is lost because the bandwidth is needed for the next refresh cycle. Similarly, there is no time to establish a communication connection for each data transfer, so connectionless services are used to avoid the overhead of setting up a connection and confirming receipt of a data message. These terms will be further explained in the Communications chapter. The important point here is that communication at this level is not at all like the client–server communication with a web page, or a TCP moderated file data transfer.

Communication protocols at this level tend to be proprietary, in spite of efforts to standardize them. Product differentiation is too important to vendors, and speed is too important to users, to freeze the technology with a standard. There are de facto standards that enable some interoperability among manufacturers, but the number of them makes the Communications chapter the largest in this book.

Second Level   Communication at the second level is between computers at the first and second control levels. The extreme reaction time requirements of the first level no longer apply, although time is still of the essence. Some proprietary protocols continue to exist, but Ethernet as a physical layer and the Internet protocols (TCP/IP) are far more common. Ethernet and TCP/IP are common because they are used everywhere, driving down the cost of hardware and software required for communication.

Computers at the second level mainly act as human-machine interfaces (HMI) to allow operators to monitor and affect the operation of controllers in the lower level. Communication may be handled as second level clients and first level servers, or a second level computer may subscribe to a group of process values, depending on the application that is running, using a temporary connectionless service. Computers that compile a process history may also subscribe to process values and alarm generators.

Fluid processes typically have graphic displays built as schematics of parts of the process, with sensor readings displayed near their actual location in the process. The values may indicate their status with regard to alarms and health. An operator may select a control point in order to operate its controller. Alarms are managed to make them meaningful to the operator. Batch processes and procedural controllers have to be able to display the current state of the procedure (perhaps by highlighting it in a map of all the states) and possible changes of state.

Discrete process HMI displays seldom show the values for all of the sensors, because there are so many of them, and because a person could only get in trouble by remotely forcing individual actuators. What is important is the state of each controlled part of the process. The operator may be offered a way to change the state, or it may only be possible to change the state of an entire machine. An operator cannot operate a CNC machine as if it were a manual lathe, but can tell the machine to stop what it is doing and eject the part.

Discrete processes tend to have many interlocks that prevent motions from exceeding their design range. Interlocks are typically a series of contacts in a Boolean AND series. In the days before computers, an electrician would approach a stopped machine with a ladder diagram and a way to tell if a contact was open or closed. Now it is possible to show the operator a list of interlock conditions and highlight those that are preventing operation. All of this requires communication with the lower level controllers.

Third Level   The third level contains the networks necessary for the computers that aid manufacturing management above the supervisory level. There are applications for asset management, equipment effectiveness, process efficiency, and so on that aid management of an entire manufacturing division at the level of people who know how the processes work. The set of levels one to three contains the entire body of knowledge of how to make the products that are sold by the company.

Fourth Level   The fourth level contains the networks required by the business systems. Business systems want to know what the manufacturing systems are doing and how they could be pushed to do more, just as managers communicated with foremen before computers. Response times are not as critical as they are for controlling processes. There is plenty of time to retry a failed communication.

It is necessary to filter the traffic between business and manufacturing. Business queries must not be able to overload manufacturing control networks, causing control messages to slow down or be lost. Business computers must not be able to operate control systems in any way, because business users are not trained in process safety. You wouldn't want someone in another city to be able to take control of your automobile with you in it.

Security   The fact that digital messages are being interpreted by application programs in computers makes it possible for a malicious message to be interpreted in a way that the automation programmer had not intended, and do damage to the program or to controlled hardware. A message can also cause itself to spread to other computers. Such messages require knowledge of the applications and the operating systems that run them. Messages can be introduced to a computer from a communication network or physically by an external device intended to transfer programs and data, such as a flash drive or optical disk.

Commercial operating systems and applications such as browsers are in wide use because they are inexpensive, but their characteristics are well known to programmers. This is necessary because one operating system vendor can not make all of the possible applications that would run on that operating system. Add to that the incentives for using the Internet to gather data about users from their computers and you have insecure systems. Automatic vendor updates leave a gigantic hole in security. The solution may require operating systems built specifically for manufacturing control systems that operate with a high level of message security and that make it difficult to physically load from external devices without proof of identity.

Autonomous Operation

True autonomy may be beyond the reach of present technology, but it might be possible to protect a process from foreseen accidents with the necessary set of computer applications and additional sensors. Japanese inventor Sakichi Toyoda created the concept of autonomation in 1902, when he added the capability of sensing broken threads in his mechanized looms. A broken thread would stop the loom so that it didn't continue to produce faulty product, as described in the Prolog. It is a far stretch from a broken thread in a loom to a refinery, but the concept does apply.

We want something more than a safety shutdown system that acts when it is too late to recover from a bad situation. Today we rely on operators who are there for other reasons, as discussed in the chapter on operators, to detect situations and correct them before damage can occur. The history of accidents shows that operators can be misled by a combination of faults into doing something that makes the situation worse. The 1979 incident at the Three Mile Island nuclear power plant near Middletown, Pennsylvania in the USA comes to mind. A hydrogen bubble formed in the reactor because the operators had the wrong model for what had happened.

An autonomous system will need more than just the process sensors because it must also be aware of its environment. Sensors are needed to detect failures of material to be in its proper place:

Vision systems to look for leaks and spills, gas clouds, smoke, material out of place, change of shape of vessels.

Microphones for gas and steam leaks, unusual noises, vibrations, explosions.

Fluid and vapor chemical sensors for wrong material or composition, or hazardous gas concentrations.

Sensors for fluid in gas pipes where it shouldn't be, such as vent flare lines.

Infrared vision of bearings of rotating machinery or vessels containing combustion.

Sense change of pressure in a building or vessel for loss of containment,

Use mass and energy balances to detect leaks and fouling.

Awareness of weather and earthquake events that could threaten processes.

Much of the work necessary to come up with situations and ways to avoid them is done when a hazard and operability (HAZOP) analysis is performed prior to commissioning a process. This requires a team of people who have the details of the process design and the experience to know what might go wrong. They are the people who should determine the use of process alarms. They gained experience by operating processes (a source of experience that may go away) and by forensic study of accidents (a source that won't go away).

When an alarm occurs, its cause must be determined, consequences evaluated, and action taken to save the product, process equipment, or lives (although there won't be any lives in an autonomous facility). There are many systems in use for working either forward from a cause with Failure Mode and Effects Analysis (FMEA) or backward with Fault Tree Analysis (FTA) from an event, but their expense has limited them to the military, NASA, and nuclear power.

A committee has to come up with every event that could possibly happen, and discover everything that could possibly cause it, and then has to either assign a probability of that happening or assign a consequence of it happening so that risk management can be applied. Risk management looks at the cost versus probability and decides if the event is worth detecting. The useful lower limit on risk is set by the 10^{-14} probability per hour of exposure that a significant asteroid will impact our planet. That works out to about 10^{-8} per human lifetime, still less than the 10^{-5} probability of being hit by lightning. On the other hand, the probability of human extinction is about 0.1 over the next hundred years (Matheny, 2007).

It appears that if you want to turn the lights out in your manufacturing facility, you will have a very large up-front expense to design the machines and their algorithms that would make that almost possible. Almost possible, because people will still be needed for preventive and reactive maintenance. Add to that the fact that one such machine will not be enough, as it represents a single point of failure. Two or three machines will be needed, each with is own immortal power supply and independent communication channel to the process control systems. They will need some way to decide which one of them is right. There is a reason why aircraft have copilots.

Human maintenance contractors will have to take care of maintenance items like filters, bearings, fuses, and stuck valves. They may be spooked by an unmanned plant, so the machines that operate the plant should have avatars with some witty dialog. Humans can reach the avatar by connecting to the plant's firewalled Wi-Fi network port or the public telephone network. A human would have to do this to gain access to the plant, just as permission to enter an operating area would be requested from a human operator. Avatars can be made quite life-like today, and that should improve with more computing power. Humans will become used to talking to avatars on mobile video phones.

Mobile robots may be needed to wash floors and surfaces, paint surfaces, detect and prevent corrosion, and extinguish small fires. Some defense against people who have broken into the plant may be required. At some point an outside group of humans may need to be called in. These outside groups will need to have enough work to stay in business, or be maintained by the government.

A conscious, self-aware machine may be required to do these things, because today's machines cannot learn from experience. There is another flurry of activity towards learning in the neural network community but they have been way ahead of available technology for fifty years. Another possibility is the genetic algorithm, which self-evolves to solve a problem presented to its inputs. Well, not entirely by itself because a human is required to provide the environment of unnatural selection that directs progress towards a goal.

Conscious machines are discussed in detail in Appendix B Artificial Intelligence.

Robots

Robots originated in fiction, going back to 1920 in Rossum's Universal Robots. The idea of manufactured servants is much older. The field is much too broad for this book, so it will be restricted here to industrial robots, defined as machines with fixed bases which can move an "end effector" to a point in three dimensions. The end effector is a mechanism at the end of the robot arm that can hold a tool or other object. If the end effector is built like a wrist, three more degrees of freedom are possible—yaw, roll, and pitch. The industrial robot first appeared as the Unimate by Unimation after the company was founded in 1956 in the USA. Japan now holds the lead in industrial (and other) robots.

The requirement for a fixed base makes the robot resemble an arm, with shoulder, elbow, and wrist joints. Each joint has limitations on forces available for acceleration and braking, velocity, range of movement, static force, dynamic position error (due to bending while changing speed), and static position error. As always, the position error has components of accuracy and precision. Precision determines the ultimate accuracy if the gain of the servo controller is high enough to reduce error to less than the limit of precision. The cost of the robot is determined by design decisions for all of these factors, and the technology available to the manufacturer.

The industrial robot, then, is a machine with controls for position in up to six axes. There is no attempt to make it look human. See http://www. tmrobotics.co.uk/6-axis.html for an example. The position controls are commanded by a computer that calculates the motions required to move from one position to another, within the physical limitations of the device. Like a biological arm, no joint can rotate more than 360 degrees, if that far. A motion command that causes a joint to move past its stops is not permitted by the computer. The robot computer, in turn, is commanded by a cell controller to make the appropriate position changes at the required times.

There are at least two ways to program industrial robots. The easiest way involves no programming. The robot is put into a learning mode such that it does not energize its joints except to cancel gravity and prevent harm to the robot. A human leads the end effector through the required motions and the program is saved. This method is not precise, but it is often adequate for the purpose. The other method is to specify positions in three or six axes for each movement. This method is more suitable to another computer that is working from a CAD drawing.

Discrete processes use industrial robots to pick up a part from one place or machine and move it to another. The part may be a ladle of molten metal or a tiny surface mount component on a printed circuit board. Many kinds of parts are assembled by spot welding, especially automotive frames, and articulated robotic arms ending in spot welders may be seen on any automotive assembly line. Similarly, articulated arms wielding paint spray guns may be seen painting complex contours on agricultural and construction machines.

You would not expect to find these robots in a continuous fluid process, where the motion of the fluids is completely determined by piping and valves. Batch fluid processes may require the addition of small to medium amounts of solid ingredients through a hatch on the reactor. There is room for error in choosing the correct ingredient and introducing it properly to the reaction that could be eliminated by a pick-and-place robot on a mobile base. This would not be cost-effective today, but perhaps it would be tomorrow. That would add a new dimension to batch control systems.

Baxter

The one-armed robots discussed above have no awareness of obstructions such as humans. They are enclosed in protective cages where they can swing within their range of motion without fear of hitting anybody. The field was open for someone to work outside those limits. A company called, interestingly enough, Rethink Robotics introduced a robot designed to work with people without protective cages. The company was formed in 2008 by Rodney Brooks of MIT with the goal of producing a robot that could work with people without trying to be a person. See http://www.rethinkrobotics.com/index.php/products/baxter/ for much more information.

Baxter does not look like a human, but it has a screen that displays two expressive eyes. The screen can rotate about the "head" to show where Baxter will move next, and the eyes (especially eyebrows) can tell a human whether things are normal or not. Baxter has two arms that have no pinch points that could hurt a human. It is primarily a pick-and-place robot, adapted to moving work pieces from one place or conveyor to another.

All programming is done by grasping one of Baxter's wrists and taking it through the required motion. Grasping closes a switch that puts the arm into zero-gravity mode so that it will not fall by its own weight. Accuracy of about 1/4 inch (5 mm) is possible, enough to pick and drop but not to close the lid of a box. An external computer is not required, so there is a limited communication interface. During programming, Baxter's screen offers touch selection of choices and the eyes express reactions to the programming. When the wrist is released, the screen nods to signify acceptance of the program.

The robots are designed to sell, priced at about one fifth of the price of an industrial robot arm. Return on investment occurs in less than six months. They are intended to act as helpers for human employees, doing the repetitive work that drives people crazy. Employees can reprogram Baxter if they think of a better way to do the job. Rethink Robotics slogan is, "It's not what your robot can do, it's what you can do with your robot."

The Dark Side of Automation

When used as a noun, automation means a set of technologies used to animate equipment. When used as an epithet, it means an implacable force that has made a major change to someone's life. Humans deal with implacable forces, such as those provided by Nature, by applying magical thinking. Magical thinking evolved as brains evolved, to prevent humans from feeling so overwhelmed that they give up on life. See "The 7 Laws of Magical Thinking" (Hutson, 2012). This is not meant to belittle the loss of a job, but to suggest that people have a way to survive that life-changing loss. The stress of change and adaptation is enormous, but of no concern to those who are focused on the money, perhaps due to a complete lack of empathy.

Jobs will be lost, and not just because machines are less trouble or less expensive than people. Consider Moore's implacable law, that the number of transistors on a chip will double every two years. Products are shrinking beyond a person's ability to assemble them. Seventy five years ago, you could strip an automobile engine down to its component parts in half an hour with a set of speed wrenches. Now the engine can only be serviced by a trained specialist. Today's automobiles are designed to be assembled by robots making over 4000 welds on the body, because the demand for cars is higher that what humans can produce.

The human drive for a small number of large corporations that manufacture everything has taken manufacturing beyond human scales. Increasing automation is the only way to meet production goals. There is no going back to human scale production because it would cost too much. This will lead to more people than jobs, with unfortunate results that are predicted by history. The economy begins to fall apart when the loss of jobs reduces the rate of consumption of products, causing the loss of more jobs.

An Early Prediction

One of the first novels to challenge the Utopian visions of promoters of automation (we'll have so much more free time) was "Player Piano" (Vonnegut 1952). Vonnegut had worked at GE after World War II and saw how some of this was going to play out.

The novel is set ten years after the conclusion of the third world war. Automation was further developed to replace the men and women who went to war. Machines now do almost all of the work, and control what a person will be assigned to do, based on IQ (remember, this was written in 1952). If you can't be a manager or engineer, you are assigned to the Reconstruction & Reclamation Corps. The people in the Corps derisively call themselves Reeks & Wrecks, indicating that there's not much job satisfaction there.

The protagonist, Dr. Paul Proteus, enters a part of a refrigerator assembly building that Edison once knew, now filled with insulation braiding and welding machines. He wishes Edison could have been with him to see the welders work as two steel plates are picked from a pile and placed in the welder. As Vonnegut describes it:

> The welding heads dropped, sputtered, and rose. A battery of electric eyes balefully studied the union of the two plates ... and the welded plates skittered down another chute into the jaws of the punch-press group in the basement. Every 17 seconds, each of the twelve machines in the group completed the cycle.

Later, as Paul wanders through the building:

> Looking the length of Building 58, Paul had the impression of a great gymnasium, where countless squads practiced precision calisthenics—bobbing, spinning, leaping, thrusting, waving. . . . This much of the new era Paul loved: the machines themselves were entertaining and delightful.

The master computer for production is named EPICAC XIV (see ipecac syrup). As a vacuum tube computer (there are limits to prognostication), it filled a large part of Carlsbad Caverns. It controlled the US economy by sending orders to human managers at plants, with no direct connection to machine groups. The machines were controlled by tape recordings of the best operators doing their jobs for the last time.

When Paul crosses a river to leave the upper class part of town to buy some whiskey for a friend, he meets one of the men who made a recording for the factory, and is now in the Corps. The conversation deepens his unease about the social issues. Perhaps he has more empathy than most managers and engineers. His unease worsens when a college friend is made redundant, making him unemployable in the "eyes" of EPICAC.

The plot thickens and Paul becomes involved with a group that wants to overthrow EPICAC by destroying factories all over the country, starting with Paul's factory. The group succeeds in stopping production at the factory, but EPICAC remains safe deep in Carlsbad Caverns and now there really is no work. The book ends with Paul marveling at the fact that men of the Corps are now applying their native abilities to repair broken machines at the factory, because that's what they do.

No happy ending there. The miracle is that they survived World War III. The message seems to be that some people are born to work with machines, even if they put themselves out of work. The book says nothing about English majors and liberal arts students beyond Paul's wife Anita, who only wanted material goods. Did you identify with Paul's fascination with the dance of the machines? You folks with fluid processes wouldn't understand.

References

Endsley, M. R. Automation and Situation Awareness. In R. Parasuraman & M. Mouloua (Eds.) *Automation and human performance: Theory and applications* (pp. 163–181), Mahwah, NJ, Lawrence Erlbaum, 1996.

Hutson, Matthew. *The 7 Laws of Magical Thinking: How Irrational Beliefs Keep Us Happy, Healthy, and Sane.* Hudson Street Press, 2012.

Matheny, Jason G. Reducing the Risk of Human Extinction. Risk Analysis, 27(5): 1335–1344, 2007, http://jgmatheny.org/matheny_extinction_risk.htm

Merriam Webster's Collegiate Dictionary, Tenth Edition, 1996.

Norman, D. A. The 'problem' with automation: inappropriate feedback and interaction, not over-automation. Philosophical Transactions of the Royal Society of London, B 327, 585–593, 1990.

Vonnegut, Kurt. *Player Piano.* The Dial Press, 1999, original 1952.

CHAPTER 7

Communications

What hath God wrought.
> —Samuel F. B. Morse (by telegraph over 57 km of wire, 1844)

Mr. Watson—come here—I want to see you.
> —Alexander G. Bell (with an acid-filled microphone, 1876)

That's one small step for a man, one giant leap for mankind.
> —Neil Armstrong (from the Moon, 1969)

Introduction

Communication is a process that uses a system of symbols, signs, or behavior to transfer information as a message from a sender to one or more receivers. Information is transferred only if both sides have the same understanding of the symbols, signs, or behavior. Protocols are used to avoid misunderstanding, especially if the consequences of misunderstanding could cause great harm, such as war.

Humans developed ways of communication before there were languages. Some ways became instinctive, carried in the genes, because we can still understand universal facial expressions and body language that affected our survival. Humans then walked out of Africa tens of thousands of years ago to grow tribes and nations in areas separated by natural boundaries such as rivers, mountains, and oceans. We all had similar brain functions, built by our genes, but there was no gene for a specific language. Now there are a great many spoken and written languages. The difference between one language and another becomes more pronounced with the difficulty of travel between the origins, as Japanese is quite different from French and Icelandic is different from Danish, although Iceland was settled by Danes. Switzerland is a nation separated by mountains into three regions with three official languages.

Automation also has regions separated by boundaries of technology and use. There are two such separations that are equivalent to opposite sides of the world. The first is between processes that are dominated by analog measurements such as flow, pressure, and temperature (fluid processes) and those dominated by two-state switches, indicators, and solenoid valves (discrete processes). The second is between field devices with microprocessors and the general-purpose computers used for control, display, and process history. Within these regions are companies that have a common basis in available technology instead of genes, but are separated by competition. You could say that a third gulf separates manufacturing systems and business systems, but that's outside the scope of this chapter.

Communications refers to systems for conveying the messages of communication, such as telephone or telegraph. Messages vary in size, type, priority, and delivery time constraints, which create different design requirements for communications systems. This chapter will narrow the scope to systems for the communication of binary messages among digital computers for process control, except for a last section on human–computer communication. It will be clear that the development of binary messages mirrors the development of human languages. There is no standard language for all humans or for all regions of process control.

This chapter begins with a discussion of the shapes of networks, the meaning of nodes, and the concept of node addresses to locate devices on a network. Then protocols are introduced as formal ways of talking to nodes on networks, starting with the Internet protocols and the Open Systems Interconnect seven-layer model, then the IEEE 802 standards.

Narrowing the focus to fluid process control protocols brings up smart transmitters, the ill-fated Manufacturing Automation Protocol, ISA SP-50, and the Fieldbus Foundation, with discussions of fiber optics, wireless, and security.

Shifting the focus to discrete process networks and protocols we find eight different but interesting protocols that each have a large user base, making it difficult to consolidate them. There are many more that are not mentioned.

Connecting diverse protocols requires gateway devices. The best-known gateways are built to the designs of the OPC Foundation and its Unified Architecture. The high points are summarized here, and the reader is directed to the OPC web site for further details.

Finally, there is a discussion of ways and means to communicate between computers and humans.

Networks

Computers are most useful when they can exchange information with other computers as messages containing data or commands and responses. Two or more computers or other special purpose digital devices connected together to exchange information constitute a network. The digital devices are called nodes, after the swellings on tree branches that can be the source of other branches. The connections between nodes are called links, which can be wired or wireless or fiber optic. Networks have a topology that describes the arrangement of links and nodes. Some topologies are common enough to have names, such as point-to-point, bus, star, ring, tree, and mesh. See Figure 7.1. If the topology is too complex or unknown, it is said to be a cloud.

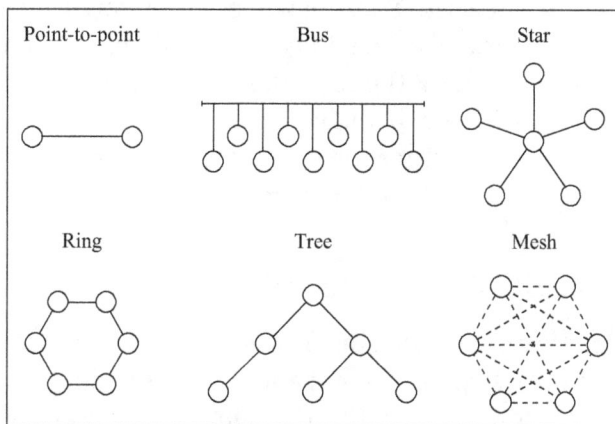

Figure 7.1. Network topologies.

The flow of information in a digital network is a series of binary bits spaced by precise intervals in time. It is too expensive to transfer bits in parallel, especially now that very high speed serial devices are available. The fundamental unit of transported information as binary bits is called a packet. The number of bits in a packet varies, but usually has an upper limit. The fundamental unit of a packet is called an octet (byte) containing eight bits. A packet has a header of information attached to the data to be transmitted. This header contains addresses and other data that is used by the network. Packets are less than 1500 octets in the Ethernet design, which take less than 2 milliseconds to send at the original network speed. This was a good match for the speed of microprocessors of that era. The network transmission speed increased at about half the rate that Moore's law predicted for microprocessors, and now exceeds a gigabit per second, sending a 1500 octet packet in 20 microseconds.

There are just two basic types of networks: circuit-switched and packet-switched. The telephone network before computers was literally circuit-switched. Operators at a central office were automated by special rotary switches (Strowger, patented 1891) that made an electrical connection to the called party such that the current that flowed through your carbon microphone also flowed through the ear-piece at the other end, and vice-versa. Now that computers do the switching, circuit-switched means a dedicated connection that is not shared with anyone else or used for any other purpose. This is not efficient for large numbers of users.

Packet-switching shares physical connections the same way a package delivery truck carries packages to many locations. The data to be communicated is divided into relatively small packets, and each packet carries the address of the recipient and the return address. The packets are multiplexed onto a high-capacity physical communication medium. The receiver assembles the original data from the packets.

Circuit-switching has a guaranteed network capacity at the cost of times of idle capacity. Packet-switched networks strive to add capacity to keep the network from choking, a technique that has worked so far. Following Parkinson's Law, the size of messages for voice, video, and spam has increased to nearly fill the available bandwidth. Soon we will discover that bandwidth is also a limited resource. Messages will not hit the wall and stop—they will slow down, with problems for services like Voice over IP and streaming video.

Network Node Addresses

Any message must have a sender and an intended receiver. Long before computers, mail was sent and delivered with two addresses: "From" and "To." Both addresses identify the name of a person, some local location identifier, and a national identifier. So far, it has not been necessary to identify the planet or solar system. As the volume of mail increased and automatic sorting and routing machines became required, postal codes were introduced to speed the routing to different areas of a country. The local "To" address is still required for the post person who actually delivers the mail. The "From" address is required to be able to notify the sender of a delivery problem. This scheme scaled up into the digital era quite well.

Connection to a network is made with a network interface device that fits the chosen media. It has an associated media access control (MAC) address which is unique on the local network. The size of the MAC address is 48 bits, with 24 bits for the manufacturer and 24 for the device type and serial number. Manufacturers usually assign a MAC address when a device is manufactured, but it may be altered by a user in some cases. Two to the 24th power is about 16 million, which seems like enough space for vendors and devices.

For the Internet, there is a second address associated with each node that is unique on Earth, called the Internet Protocol (IP) address. It has 32 bits grouped as four octets and written as four groups of decimal numbers, such as 192.168.42.1. The address defines a local network and an extended network. There are three ways of doing this, defined as three classes for small, medium, and large local networks. See Figure 7.2 for the arrangement. Class A networks can have about 16 million nodes but only 127 networks. Class B has about 65 thousand nodes and networks, while class C has only 255 nodes but 16 million home and small office networks.

These addresses locate a specific node, but sometimes it is useful to address a message to all nodes on the network. This is called a broadcast. An example is a broadcast message asking who has a specified IP address. Only one device will reply, if any, so the network will not be flooded with replies. The reply is used to learn the MAC address associated with that IP address. A broadcast can also be used to locate a device that provides a specific service.

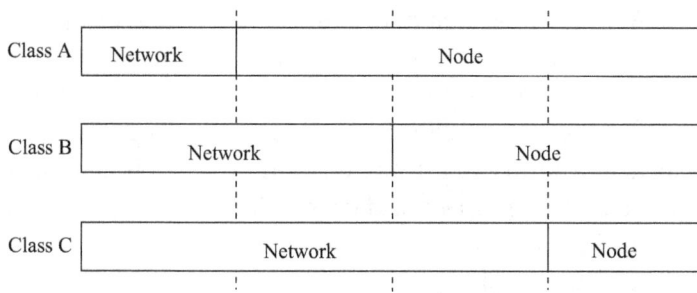

Figure 7.2. IP addresses.

Node zero is not assigned to a device because it is used to indicate a broadcast. The MAC broadcast address has all bits set to True.

When the internet protocols were being developed, 4.3×10^9 addresses seemed like enough to get the project rolling. Everyone knew that it would be painful to change the address structure, but that's what they did to end a year's debate on address schemes and get started on the main project. The three-class system described above was used for version four, known as IPv4. Only 14% of the possible addresses have been used, but the last block of class A addresses was allocated in February of 2011. Work beginning in 1992 or so, as the addresses ran out, resulted in version 6, or IPv6.

IPv6 has 3.4×10^{38} addresses in 128 bits or 16 octets instead of the four octets of IPv4. This is not enough to address each of the 10^{79} atoms estimated to be in the universe, but it is enough for 10^{28} addresses per known intelligent life-form at this time. Truly, this seems like enough.

IPv6 was launched in June of 2012. It is not interoperable with IPv4, but it is convertible. The basic packet header of IPv4 was changed to improve routing in a large system. The packet size limit was greatly increased, and changes were made to improve privacy. The details are beyond the scope of this book. Many LANs will continue to use IPv4 while new network routers at Internet Service Providers (ISP) convert to IPv6 for compatibility with the cloud.

Local and Wide Area Networks

A local computer network (local area network or LAN) originally meant a set of computers that were connected to a common physical medium. The use of switches, repeaters, and wireless have eliminated the need for a common physical medium, so now a local network consists of nodes that have the same network address prefix. All MAC addresses still have to be unique.

A wide area network (WAN) interposes a routing device between the local network and a network consisting of computers whose only purpose is to route packets from one network address to another. The node address is sorted out by a router at the other end. In the beginning (1985) this network was slower than a LAN, but now it is many times faster as fiber-optic connections and purpose-built routers respond to the demand for more capacity.

In the early days, drawings of networks enclosed the details of the WAN routers in a fluffy cloud. That cloud has a new meaning today. Some people speak of the cloud as a source of infinite wisdom, when it is just machines on the Internet that rent out storage space and processing time. Figure 7.3 shows nodes in a cloud of unknown connections accessed by nodes outside of the cloud. The security of your data is no better than with any other Internet transaction, and you have sent it to a vendor who may or may not be protecting your data. A cloud is also a symbol for uncertainty, which is not what most people want when controlling a process.

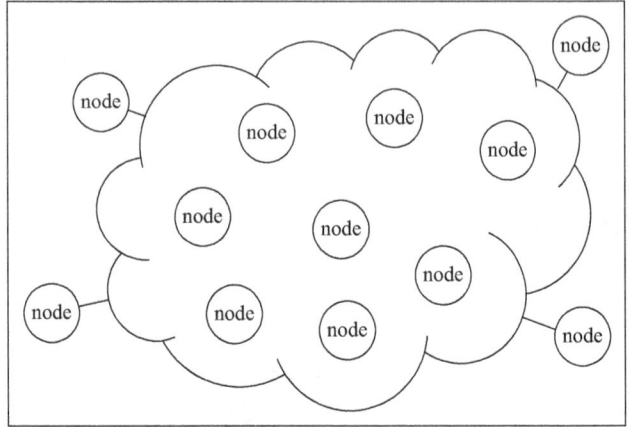

Figure 7.3. A network cloud.

Normally, a WAN would have to use the same address scheme that all of the connecting LANs use. The use of a truly universal address scheme makes it possible to use other schemes in the WAN, because the WAN carries packages of bits from one portal to another without examining them, even as aircraft carry pods of packages as air freight. A universal address is unique within its universe of possible delivery sites. When a router delivers a LAN packet to the WAN, it must use a destination address that can be interpreted in the WAN. The contents of the packet, including the local MAC address, do not matter. Similarly, an air parcel delivery system takes packages with city and state addresses and consolidates them into aircraft cargo containers by airport in a nearby city in the same state. The contents of the cargo containers are then sorted by city and placed in delivery trucks, perhaps bound for further sorting until delivered to the package address.

Protocols

Diplomatic protocols are used by international ambassadors to avoid giving unintentional offense while communicating with people from different cultures. It is essential that nothing be misunderstood. Communication protocols between organizations define the rules for communicating by written words, such as the proper titles for members of the organizations.

Digital communications protocols are sets of rules that allow strings of binary bits to be passed among machines in a way that the contents can be delivered and understood. Communicating computers probably had different human programmers who would not use the same algorithms for communication unless they were working from a common set of programming standards. There are a great many programming standards, so the ones used for digital communications are called communications protocols. As someone once said, "Standards are good because there are so many of them to choose from."

A digital communications protocol for computer networks is an application program that sits between whatever physical medium that carries network information and a user's application program. It is similar to the computer operating system that sits between the computer's peripheral devices and an application program, except that an OS also provides control of program execution and memory management for the user's application. The protocol can be quite complex, and so the problem of designing a protocol needs to be subdivided into smaller tasks called protocol layers.

It is true that the user does not have to be aware of the complexity of the network that transmits data from one user to another. A user with a cell phone does not care about the mighty technological developments that underlie the caller's ability to talk to someone anywhere else on Earth that there is coverage. We will scratch the surface of protocols to show the differences between a transaction-oriented network like the Internet and the protocols that are necessary to support process control.

The Internet Protocols

One of the first high speed networks developed was Ethernet, so named because it worked in the various "ethers" (also spelled aether) of cable, telephone, and radio. It was developed at Xerox PARC by Robert Metcalf and others, first described in 1973. It competed with proprietary technologies, but it was open and less expensive so it became the dominant technology in a decade.

The Defense Advanced Research Projects Agency (DARPA) developed packet-switching into the architecture and protocols that are the foundation of the Internet during the late seventies. The architecture is based on a simple three-level hierarchy that commonly sits on the Ethernet media access control (MAC) and physical layer (PHY). See Figure 7.4. The foundation is a packet delivery service called the Internet Protocol (IP). Below it is any physical network that can switch and deliver packets, such as Ethernet. IP uses a number of protocols to deliver a standard data packet from any host computer to any destination computer that can be reached by any route that can deliver the packet through the cloud of gateways and routers.

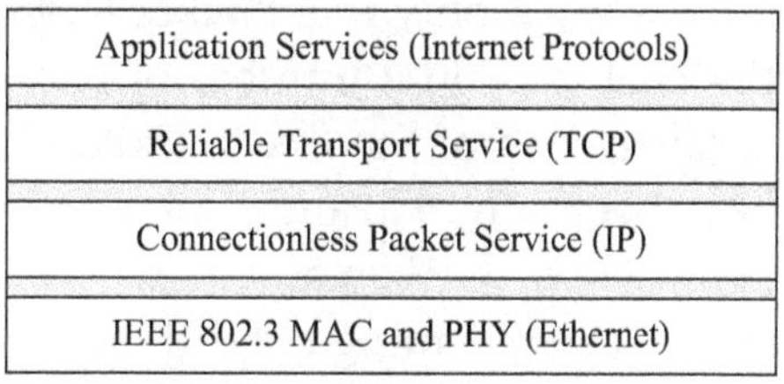

| Application Services (Internet Protocols) |
| Reliable Transport Service (TCP) |
| Connectionless Packet Service (IP) |
| IEEE 802.3 MAC and PHY (Ethernet) |

Figure 7.4. Internet protocol layers.

The packet has a standard header of 20 octets that is used for all IP packets regardless of the size of the appended data. Since the physical networks that may carry the packet have varying limits on the maximum size packet they may carry, IP can use fields in the header to fragment the original packet into smaller packets that all have the same host message identification number. The fragments may take different routes through the cloud, which may have different transit times or get lost altogether. IP is an unreliable delivery service, but one that makes a best effort to deliver packets anywhere within the scope of the Internet. The best effort is not trivial because there are Internet routes that reach most of the planet, including those that use communications satellites.

The second layer of the Internet architecture provides reliable transportation, primarily by collecting packets until a complete message can be reconstructed. This could be done by individual user applications, but network reliability depends on doing it right, which is not guaranteed with many different versions of doing it. DARPA developed the transmission control protocol (TCP) to provide one common solution to the problem. TCP is commonly associated with IP as TCP/IP, where the slash should be read as "over" because TCP alone can be used over any unreliable packet delivery service.

TCP requires a full duplex connection between computers similar to a circuit-switched connection between two phone numbers. An entire IP address can't be tied up for this purpose so the concept of "ports" was introduced. Each IP address may have 64,000 ports, which seems like enough for one computer. The first 255 ports are reserved for Internet protocol applications. For example, two ports are reserved for network bootstrap programs, port 80 is used for hypertext transfer Protocol (HTTP), and port 123 is reserved for the network time protocol (NTP), which can synchronize machines within milliseconds over a packet-switched network with random delays. There is a connectionless and unreliable user datagram protocol (UDP) that is independent of TCP, but uses the port addresses to connect to specific applications. In general, applications that use UDP can detect loss of packets by timing the expected UDP reply packet. UDP is not suitable for large messages.

It is not possible to have reliable transfers without having the destination confirm receipt of each packet back to the source, which is the reason for the full duplex connection. If the source does not receive confirmation in a preset time, it resends the message and starts the timer again. Considerable time can be wasted if only one packet is sent at a time. The duplex connection allows the host to send a number of packets, perhaps eight. When the host receives confirmation of the first packet, it sends the ninth and so on until all packets have been sent. The only delay is the time spent waiting for confirmation of the last packet. TCP can adjust the number of unconfirmed packets in flight to suit the network loading conditions.

The third layer of the Internet architecture contains the application services that interact with user applications or with Internet applications that maintain the network, such as router reconfiguration when bottlenecks or losses occur.

The OSI Protocol Model

The Open Systems Interconnect (OSI) group of the International Organization for Standardization (ISO) began work on a layered model in 1977. The idea for the standard was proposed by a group in the UK representing industries that were concerned by their predictions for larger networks of more powerful computers. The committee settled on seven layers, and so the communications world was buzzing with news of the seven layer model from about 1980. The model was considered to be too complex by some computer scientists, notably Andrew S. Tanenbaum, giving credence to the saying that an elephant is a mouse designed by a committee. There were so many optional features that vendors using the standard could build products that would not interoperate with other vendor's products.

General Motors (GM), as a large user of automation products, developed a communications standard called the Manufacturing Automation Protocol (MAP) to allow the products to interoperate. At the time, Ethernet was still young and widely believed to be unsuitable for process control because nodes could begin transmission at any time. The result on large networks was time lost while recovering from collisions. GM chose the OSI model as the only one that would scale up to their needs. They must have limited the options to a profile that fit their needs in order to assure interoperability. MAP was released in 1982, allowing vendors who wanted to do business with GM to develop products to the standard. Demonstrations of interoperability among 21 vendors were held in 1985.

Meanwhile, Boeing had the problem of networking the workstations used by designers of aircraft parts. They developed a Technical Office Protocol (TOP) and merged it with MAP in 1986, calling the product MAP/TOP. Further complicating things was a series of revisions to MAP leading to MAP 3.0 in 1987, which were not entirely interoperable with older versions. That is the way with new and complex projects, but as MAP settled down, the rest of industry adopted Ethernet. The IEEE committee 802.4 for MAP was disbanded in 2004 due to lack of attention.

The OSI model was ahead of its time. Computational power was not adequate to process a seven-layer protocol in a timely manner. One user who had attended the 1985 demonstration reported that a command from an operator might take eight seconds as the command traversed layers up and down as it went around the ring on its way to the destination. The time spent on Ethernet collisions was insignificant compared to that.

MAP may be gone, but the use of protocol layers arranged in a stack has survived. Now that computers are faster, the time penalty for traversing layers is not significant. The advantage of having multiple design teams working on the layers after agreeing on the interfaces between layers is considerable in a large project. So is the advantage of dividing standards committees into subcommittees for each layer.

The OSI Stack   The seven-layer stack of protocols shown in Figure 7.5 begins with the first layer, at the bottom. The Physical layer works with the raw data from the movements of electrons in circuits driven by rapidly changing voltage differences in the wired medium, photons from fibers to photocells, or electromagnetic fields detected by an antenna. Each of these media has its own transducer between the physical signal and the logic levels familiar to microprocessors, in both directions for transmit and receive. The physical layer specifies the properties of the media that are common to all devices on the network, because interoperability begins with having a common idea of what the electrons are doing. The means of transducing the message in the media to logic levels is completely up to the vendor. The transducer, called a network adapter by OSI, may be an electrical modem, an optical modem, or perhaps a radio. The physical layer also specifies the physical nature of the network in terms of transmission distances, types and number of devices allowed on the network, and physical means for traffic flow control.

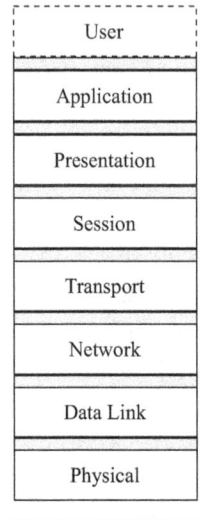

| User |
| Application |
| Presentation |
| Session |
| Transport |
| Network |
| Data Link |
| Physical |

Figure 7.5. OSI protocol stack.

Each layer has an upper and lower interface. Each interface must be interoperable with the adjacent layer. Interoperability requires that adjacent layers have exactly the same understanding of the behavior of the signals that cross the interface. There must be no undefined behavior, or behavior that is known to one layer but not the other. The physical layer interface with the communications medium depends on the vendor's design of its network adapter. The upper interface transmits and receives strings of bits that when passed through the network interface will be seen the same way by any other device on the network.

The Data Link layer handles communication on the local physical network, arranging and timing the bits to suit the physical layer. It only works with MAC addresses. It provides the services required to add a node to the network and to recover lost data link messages. It may also provide time synchronization among the local nodes if it is the designated time server, as required for scheduled control messages.

The Network layer works with network address fields to handle traffic among different networks. Routing takes place at this level. If Internet protocols are used, this layer translates the node part of an IP address to MAC for the data link layer. It provides the services necessary to operate the node as an IP device.

The Transport layer handles errors such as retries of lost messages and ordering of out-of-sequence messages, so that messages are sent to the next upper layer as they were sent, no mater how individual packets were routed.

The Session layer manages the opening, maintenance, and closing of sessions, which are connections between two computers that will carry what amounts to a discussion with periods of silence for thinking, much like a telephone call that does not come from a telemarketer.

The Presentation layer provides a uniform data interface to the user by handling encryption and translation of data formats.

The Application layer is the last layer between the protocol stack and the user applications. It provides services to the user such as read and write of resources in other machines.

The last two layers are not well specified because they depend on the application of the device containing them, and the interoperability requirements for that application. Some applications may not use the layers between Data link and Application. Routers and bridges that pass traffic between local networks may not have any layers above the network layer.

The IEEE 802 Series Standards   The IEEE started a series of standards in 1980 to define the media access and data link layer interfaces for the OSI model:

802.1—originally introductory, now bridging of separate networks into one, unlike routing

802.2—defines the upper interface of the data link layer, including protocol data units

The rest of the standards deal with media access:

802.3—Carrier Sense Multiple Access with Collision Detection (CSMA/CD) as used in Ethernet and others

802.4—token bus, actually a virtual token ring without the advantages of a physical ring, now disbanded

802.5—token ring, designed by IBM, still used for its reliability

802.6 through 802.10 have been disbanded

802.11—wireless, with many subgroups identified by letters such as b, g, and n

802.12 through 802.14 have been disbanded, although 802.13 may have been considered unlucky

802.15—personal area networks, such as Bluetooth

The list goes on, but they are not relevant in this book.

CSMA/CD refers to media access control that allows any device to start transmission whenever the network is quiet. The cable distance between nodes allows another device to start transmitting without initially detecting the other device. Collision detection then causes both nodes to back off and restart after different delays that are larger than the cable round trip time. This behavior makes the network non-deterministic because it is not certain when a message will be successfully transmitted.

Token refers to the requirement that a node possess a message that it has permission to transmit. Consider a meeting of a dozen people. Normally, they use CSMA/CD to speak when others have stopped speaking, but when the discussion grows heated a token is required. Try designating a white-board eraser as the token at your next large meeting, and announce that no one may speak unless they have the eraser token.

Peer-to-peer and master-slave are methods of controlling who will speak. A peer holding the token may pass it to anyone else when done speaking, but usually to the next node address. A slave must return the token to the master, who will decide who is next.

Process Control Communication

The Internet protocols are designed to deliver messages far and wide, using the best available methods. These methods are not good enough to meet the timing requirements for process control. Communication becomes mixed with the timely execution of various control functions that have been ordered by a control engineer to meet a specific control requirement. Digital computers have to work with samples of the process conditions, where the sampling is fast enough to assure an accurate model of the analog signal. The ordered control functions have to be executed completely after each sample. This can be once a second for fluid processes or 100 times a second for discrete processes, give or take a factor of ten.

Communication for industrial process control systems has different requirements at different levels. The controller level network, such as the data highway that connected the controllers and human–machine interfaces of the first distributed control system in 1975, requires some sort of scheduling to assure that cyclic control data always gets through. The network is shared with messages that vary in size and occurrence by using the time left over after the scheduled control messages. The device level network did not appear until microprocessors invaded field devices in the early eighties. It may carry only control messages if the cycle time is tens of milliseconds or less.

The fluid process control field was dominated by a few large companies, which limited the number of competing protocols. This was not the case for discrete control. One vendor got an early start with an open protocol that filled the network vacuum in the late seventies, but then things exploded. The reason was that the technology to support networks was not invented by control vendors but by integrated circuit manufacturers, so that it became available to all control vendors at the same time. Each vendor's R&D group then set out to develop a protocol that suited their view of the market.

These parallel developments accumulated bases of users who could not easily switch to another vendor, which was the whole point of proprietary protocols. The outcry from users for standards did sort out some winners, but the goal of a common standard proved impossible to meet. See the story of ISA SP-50 in this section.

Serial communication has been used almost as long as there have been process control computers. Vendors asserted that standard RS-232 or RS-485 was used. The problem is that RS-232 is a physical layer standard. Vendors could and did use every imaginable scheme for coding the bytes (this was before octets) in the RS-232 data stream. There was no interoperability at the application level. There was no need for interoperability within a control scheme before control applications were divided among field devices.

The 1982 MAP/TOP protocol was described above in the section on the OSI Stack. It was better than RS-232 but not suitable for control distributed among field devices. We will begin with fluid processes and then consider discrete processes.

Fluid Process Networks and Protocols

The division between fluid and discrete processes is not as sharp as it is described for the purposes of this book. Each contains elements of the other kind of field device and network. The dominant network will be the one discussed here, but the process automation system will have to be able to talk to both kinds of network.

The oldest networks linked process control computers, but we will start at the lowest level with the field devices.

Smart Transmitters

Microprocessors migrated into field control devices in order to linearize their output. The devices used the electrical 4-20 MADC (milliamperes direct current) standard, which leaves something less then 4 MA to run the electronics. It wasn't long before ways were found to modulate the DC with a data stream without affecting existing analog control devices. The first one appeared in 1983, followed shortly after by the second, each by major vendors. One of the main selling points was that a transmitter could be checked without sending a person out to climb to one of the inaccessible places where transmitters were placed by plant layout designers. Not only that, but it could have its range changed to meet changes in the process without being removed and brought back to a calibration workshop. The labor savings justified the extra cost.

Naturally, the two protocols are different, and both are still for sale today. One is called HART (Highway Addressable Remote Transmitter) and the other DE (Digitally Enhanced). HART has become an open standard, and is widely used. Improvements in hardware have increased the uses for the protocols, and improved hand-held communicator (HHC) devices for instrument technicians have increased their adoption.

The smart transmitters raised the problem of changing HHC displays to suit the field device at the other end of the wire. At first, there was only one family of pressure transmitters. The early HHC devices did not have sophisticated displays, just lines of text. A numbered menu tree, perhaps six menus deep, eventually displayed the desired information. Temperature transmitters needed a different menu. As the list grew, a device description language was developed that could be interpreted by the HHC. When larger memory chips became available, it became possible to download the HHC device with a selection of transmitters that the maintenance person would encounter.

Graphic displays added another layer of complexity, and soon it took a disk file to store the description for one field device. Then the distributed control system had to display the data and that feature added the dimension of multiple host vendors. All of this was accomplished, but the time for development allowed a competing standard to appear for the device description language, with its own large user base. So now there are at least two ways to describe field devices that a display device must handle, but at least there aren't twenty.

Digital Device Communication

The smart transmitter protocols were not truly digital because the device primarily used the electrical DC signal to transmit the process value. The fact that 4-20 MADC is a fixed range means that it defines a measurement span. Each device is calibrated so that 4 MADC equals the low range physical value and 20 MADC equals the high range value. The device is constructed with a microprocessor to maintain accuracy and linearity from the low to the high end of the span within the environmental limits of the device. True digital communication eliminates this translation and sends the numerical process value. Unfortunately, this leaves the user with no idea of what that number means in terms of the range that the process designer selected for the measurement. The concept of range is still required in a digital value system.

MAP was too complex to put in a field device with its low-power microprocessor, so vendors began looking at ways to do digital communication. IBM's success with opening the specifications for interfacing to a personal computer was not lost on the vendors, and so they realized that there had to be a common standard for communication. Another driver was the fact that transmitter manufacturers needed to have their products work with all control systems.

The ISA SP-50 Experience

Driven by the MAP work, users and vendors urged the ISA to re-open the Standards and Practices SP-50 committee on field device signaling standards. Back in 1950, the problem was to choose between existing 10-50 and 4-20 MADC spans for an electrical signaling standard to add to the 3-15 PSI pneumatic standard for signals between field instruments and control room panels. The lower current range had clear technical advantages, but agreement on 4-20 was not possible until after the principle vendor of 10-50, who had used magnetic amplifiers (as discussed in the Control chapter), had established a base of 4-20 devices with transistor amplifiers.

The field devices (analog transmitters and valves) that would be adapted to the digital world by new microprocessors and flash memories had to survive in harsh environments while sipping about 50 milliwatts of power from the same pair of wires that transmitted the signal. There was not nearly the demand for low power microprocessors that there was for desktop devices drawing hundreds of watts, so performance did not match that of the desktops. Also, there were no hard or floppy disk drives and certainly no fans. A liquid crystal display could provide a human interface of sorts, but there was nothing like a keyboard.

SP-50 was reopened in1985, but it was too late as vendors already had proprietary solutions for elementary digital communication and they were all different. Added to that was the perception that SP-50 would lead to the single standard for digital communication. It became obvious that this was not possible because of the vastly different timing requirements for discrete and analog sensors. Dr. Richard Lasher of Exxon proposed that the standard be divided into two hunks, H1 for analog and H2 for discrete (he was an analog guy). H1 would run at 31.25 kbit/s because that had been established as the maximum rate for running digital signals over existing analog wiring. Far too much had been invested in the existing analog wiring to replace it for digital signals. For example, "home run" cables from a set of field devices to the central control room were either run in overhead cable trays, where they could be destroyed by a crane or a fire, or buried in trenches. Exxon chose to bury them in trenches, and to fill the trench with red concrete so that no one digging would try to dig into red concrete. There was no way to add new cable to the trench. H2, on the other hand, could use whatever bit rate they chose.

The politics of competition was still dominant in 1988, but things calmed down a bit when it became apparent that no existing digital protocol would be adopted. Things became tense in the layer subcommittees as proponents of the system they knew best fought for supremacy. Meanwhile, Dr. Lasher started the User Layer subcommittee (SP-50.4); promoting the work that Exxon had done with Honeywell to produce TDC-2000. The user layer specified process control function blocks with specified behaviors for the parameters that were exposed to fieldbus. There was remarkable agreement among members of the subcommittee, partly because the functions were ancient history and partly because Lasher made so much sense.

There were three major goals that Lasher wanted from distributing control applications into field devices:

- Reliability—because the field devices had to have reliable hardware or they wouldn't sell, and because the software would focus on those control applications that could be thoroughly tested, unlike the multi-megabyte programs in DCS systems that had an untestable number of interacting options.

- Fault tolerance—because, as long as there was power to the fieldbus, the PID control algorithms would "keep on PID'ing" no matter what happened to the more complex network of hosts. He did not foresee that the hosts could be compromised by the likes of Stuxnet, but then, no one did.

- Low cost per loop—because the cost of software development could be spread over many more devices than host computers. The major cost of a field device is the machining required to build an indestructible housing for the sensor and its electronics, so software was never a major part of the cost.

He was right, but the low cost thing didn't come to pass. The expense for developing software and training sales engineers was not negligible.

Progress was slow, which annoyed the vendors who wanted to have a standard and get on with making devices. It seemed to them that committee members were solving the more interesting general cases, rather than focusing on specific solutions. It was also true that the problem was not going to be solved by one standard. In the end, there were eight of them.

Foundation Fieldbus

The vendors lost patience in 1992, so a consortium of vendors formed the Interoperable Systems Project with the purpose of bypassing the endless discussions of SP-50 in order to produce a useful analog fieldbus. The applicable work of SP-50 in the communication stack was adopted, but Lasher's work was rejected because it was "too complicated." It is more likely that Lasher's insistence on interchangeability doomed that work. Interchangeability would have frozen the control functions in time, which many users wanted as it would reduce training and vendor lock-in. But the vendors had control, and they wanted interoperability of devices differentiated by marketable features and benefits. The result was a user layer that was in some ways more complex than Lasher's, but drew heavily on his work with SP-50.4.

The Interoperable Systems Project reorganized as the Fieldbus Foundation. A group of talented engineers was isolated at a hotel in the Black Forest of Germany in November 1993 and told that they would get their passports back when they had a fieldbus specification. Naturally, a specification was produced in about three weeks. Just kidding about the passports. This went on to become the foundation of the Foundation Fieldbus specifications for an analog system (as opposed to discrete) fieldbus. Discrete data could be transferred with fieldbus, but the message size was unacceptable for high speed PLCs. More years went by as the early kinks were worked out of the system, and vendors learned how to interpret the specifications by participating in interoperability trials.

The Foundation developed a suite of interoperability tests that allowed them to certify a device as interoperable on a fieldbus network. This meant conformance to the specifications, but vendors were allowed to develop extensions to the specifications. This was necessary because vendor marketing departments have a mind-set that their product must be differentiable from others. In fact, users are more likely to differentiate vendors by service capability. The specifications pretty well covered what was necessary to control a process, but vendors marvelously diverged on device diagnostics. This proved impossible to ignore, so device description files were developed that let a host computer read a file from the device vendor in order to decode and display the manufacturer's maintenance data from the device.

There are many communications busses for field devices, including those in cars and aircraft. This is why SP-50 could not settle on one bus that fit all users. The members of the Fieldbus Foundation were primarily interested in the data-rich instruments used for fluid process control, with the exception of Siemens, whose managers wanted control to remain in the controllers.

Control in the Field Fluid process control mainly consists of many control loops composed of sensor, controller, and valve positioner, as shown in Figure 7.6. Some of the loops can become quite complex by adding mathematical functions to calculate the measurement to be controlled, or by using override control schemes that select the output of the controller whose measurement has passed a process constraint. Control schemes are initially designed on paper as blocks containing control functions that are connected together by links that show the direction of data flow. These are called function block diagrams.

Figure 7.6 shows three connected function blocks in two separate field devices. The analog input block contains configuration data that parameterizes a fixed analog input application program in the device operating system. Data from a process sensor (the circle marked Flow) is converted to a standard class of output parameter that can be understood by other fieldbus devices. The signal leaves the transmitter device and travels over the bus to the valve

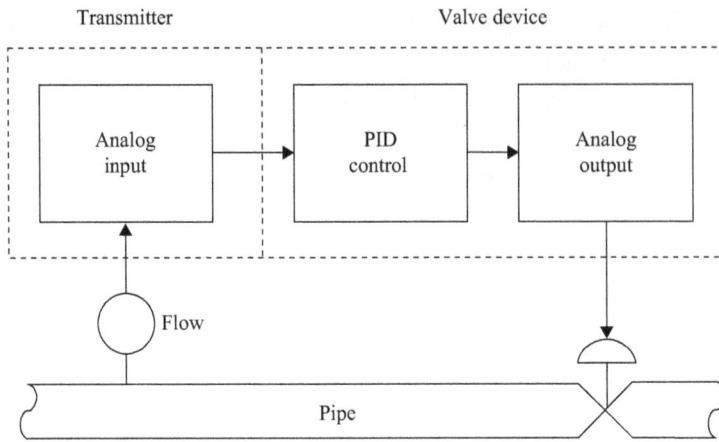

Figure 7.6. Control in the field.

device, where it is linked to the process variable input of a block representing the control function. Without going over the bus, the output of the control block is linked to the input of an analog output block. This block converts the standard fieldbus value to something that will change the position of the valve stem and the amount of flow through the valve. The three blocks must be periodically executed in order from left to right no matter where they are in physical devices.

Distributing the control loop among physical field devices that measure or manipulate process variables raises the communication problems that go with distributed computing, such that each function block that executes in a specified sequence has fresh process data for the computation. The function block chain must be executed about four times faster than the shortest time constant in the process to avoid Shannon's sample aliasing. The time constant is usually determined by the pneumatic actuator that moves the control valve, which may be in the range of one to ten seconds. There is a time limit for making the communications required for the calculations, which is set by the time to start the next cycle. If block execution and link communication fill the cycle time, there will be no time left for report or configuration messages. The situation is further constrained by the speed of the communications medium, which was 31.25 kbit/s over existing twisted pair wiring in the late eighties.

FF Protocol Layers   Lasher's user layer for SP-50 drove the communications scientists working in the lower levels to new designs. The OSI model provided names for the layers and guidelines for dividing up the work, but much of their work was inadequate for process control. Communications for distributed systems was not a new concept, but the constraints imposed by field devices were novel.

Figure 7.7 shows an approximation of the protocol levels in a fieldbus device. The bottom layer is the physical layer designed by SP-50. Almost all of the analog field devices developed for fieldbus had to use existing plant wiring so the 31.25 kbit/s physical layer was used. The SP-50

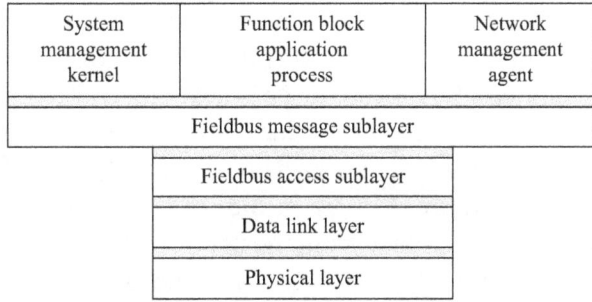

System management kernel	Function block application process	Network management agent
Fieldbus message sublayer		
	Fieldbus access sublayer	
	Data link layer	
	Physical layer	

Figure 7.7. Field device protocol stack.

data link layer is a marvel of complexity that accomplishes many functions in a small space. To aid the adoption of fieldbus, the data link layer protocol was designed into an application specific integrated circuit so that the device microprocessor would not have to handle it. The OSI application layer was divided by SP-50 into two sublayers. The fieldbus access sublayer (FAS) manages the different methods used for communication with the device. The fieldbus message sublayer (FMS) provides services such as read and write to the layers above it. The services are defined in an application interface section of the layer's specification document.

The three application processes above the fieldbus message sublayer manage aspects of the device and provide the space and services required for executing control function blocks. The system management kernel is able to run by itself when the device starts up in order to initialize the more complex applications above the data link layer. When the device has started up, system management provides services to locate data by its tag name and to synchronize the device with social (wall clock) time. Network management provides services to manage communication over the fieldbus.

Communication Objects The SP-50 application layer had developed an Object Dictionary (OD) format for the description of communications objects (from the object-oriented programming viewpoint). Each communications message is simply a bag of bits unless the scheme for encoding and decoding those bits is known. The OD provides the encoding for each object, which is located at a unique index in the OD. It is also necessary to identify the virtual field device (VFD) within the physical device because the OD varies with the data in the real or virtual device, except for initial data about the sizes and starting locations of things. Any host can read an OD to get the keys to decoding messages from the device. As it turned out, many of the object definitions became standard and so the OD does not need to be read for the definition of standard objects.

Each control function block has a set of parameters that characterize the function for its intended use. The first parameter is the unique name of the block, expressed as a string of 32 octets. Before computers, field devices were identified by stainless steel tags carrying the designation of a location in the process where the device would be used. In fieldbus, the tag name provides the only way to find a physical device from the network if its node address is unknown. Once the device is located, a path may be established by tag name to any VFD within the device, and its contents may be found by OD index.

All of the network-visible parameters in a block have unique index values in the OD, providing a numerical address for each parameter. If it isn't in the OD, it can't be seen from the network. Standard function blocks have a standard set of parameters that have OD index values offset by the same amount from the OD index of the block origin. Roughly, a VFD name corresponds to an IP address, and an OD index corresponds to a port.

Fieldbus Links   There are two kinds of links—physical links between fieldbus nodes and the logical links between function blocks. The following is about function block links.

In order to connect the function blocks into a control scheme, an output parameter in one block is linked to an input parameter in another block. An output parameter can be linked to many inputs, but each input can only come from one output. Otherwise, successive outputs in one computation cycle would overwrite the previous input value. A device handles these links with no delay in the cycle if the blocks are in the same device. If a link is external, the communication of the link data must be scheduled to occur after the source block has been executed. The execution of the block receiving the link must be scheduled to occur after the communication is complete.

There were those who thought that scheduling would not be possible, Lasher among them. Furthermore it wasn't necessary—you just ran each set of blocks before a communication four times faster than the set of blocks following the communication in another device. The error would never be more than a quarter of the slower block set's cycle, but there would be four times as many messages occupying bandwidth than if you scheduled it. Since bandwidth was at a premium at 31.25 kbit/s, scheduling won. You may be wondering what would happen if a link message got lost, because there is no time to retry it. Each function block input has an algorithm for dealing with complete loss of communication. Loss is not declared until three messages have been missed, giving a welder time to strike an arc on the metal next to the field device.

All fieldbus data connections are made with virtual communication relationships (VCR). Each VCR is defined by a table of data that is stored in the device network management data base. Each VCR contains 17 objects that are defined in the OD as communication objects. The first item uniquely identifies the VCR. The second specifies the type of communication required: publisher, subscriber, report source, report sink, client, or server.

There are three basic kinds of communication:

- Publish/subscribe is used for the process values that are linked in control function block schemes. The device containing the source of the data publishes (broadcasts) it to the bus. It is read by any device or host that has subscribed to it. This service is scheduled so that it occurs when a fresh value is computed. Subscribers maintain a buffer for the published data so that the subscriber can retrieve the data when it is ready to use it, not unlike a newspaper delivery to a subscriber's box. When the subscriber retrieves the data, it sets a flag to indicate that the data is stale (used). The next message from the publisher clears the stale flag. No confirmations of receipt are sent because there isn't enough bandwidth, and the publishing device could be overwhelmed by the traffic. Unlike a newspaper, it is not possible to request a new paper if the box is empty.

- Report source/sink is used for device to host communication, usually to report events, alarms, and small blocks of process variable data for trending. The device, as source, may send to one host, multicast to several hosts (the usual case for redundancy), or broadcast. The host sink must have a corresponding VCR for the other end of the transfer. It is not a connection. Alarm reports are confirmed with a separate client/server connection from host to device. The device will continue to resend the alarm after a delay until any host confirms reception.

- Client/server is used for communication between a host computer and a field device. The connection is similar to a telephone connection in that it can be initiated, maintained as dedicated to the source and destination that set it up, and terminated when it is not needed. Each field device has at least one VCR for a server, which allows various host clients to connect to the device in order to read and write objects in the OD.

The remaining 15 objects in the VCR identify the remote address, if any, information for the data link layer about size and timeliness, the VFD identification (to differentiate VFDs within a node address), and a bitstring (series of checkboxes) for services required from the FMS layer. If the destination device can't provide the services, the connection fails.

The VCR table described above is configured by the user (actually, by the application used to configure function block applications). This data does not change unless an external device changes it, so it is said to be static. There is also a dynamic VCR table that is not configured and is reset to zeroes whenever the device restarts. This table contains nine fields that store the states and times associated with connections. It is modified by the FMS and FAS layers. More dynamic information about connections is kept in the data link layer, but the FAS messages have been translated into data link messages that have no knowledge of VCRs. Due to the size of the VCR structure, the number of them available is limited by the low power memory used in field devices. Some user will always be able to find that limit and complain about it.

The data link layer has to provide services for scheduled communications and synchronizing the time in all of the devices on a physical link so that they will all start a computation cycle at the same time. It is able to provide services for point-to-point, multicast to groups, and broadcast to all devices.

Using a Fieldbus Network Configuration is the process of loading the various object dictionaries with data derived from a function block application that was designed by a human, using a configuration tool application running in a host. System management and network management are also configured to accomplish communication of the linked block values. Much of the linking configuration, such as VCR records, is done automatically by the tool using information from the function block configuration. The user has to specify the order of execution of the blocks, which determines the block execution and link message schedule. The result is stored in files on the host, and downloaded to the devices when a physical network is available. The devices have flash memory to remember the information through a power failure. It is possible to load each device from configuration files in the maintenance shop before installing it in the field.

Each computer or digital device on a network has to be configured with some network management information, beginning with the node address. A computer with a Windows operating system has a configuration page for TCP/IP under Properties for your local area connection. The Dynamic Host Configuration Protocol (DHCP) may be used to get address information automatically from a DHCP server on the network, which is normally the router between your local network and your internet service provider. If DHCP is not used, you must configure the IP address, the subnet mask, and the default gateway to the Internet. You will also need the IP address of a Domain Name Server (DNS) that can turn "www.momentumpress.net" into the IP address for the computer with that domain name. If DHCP is available then all you have to do to put a new computer on a network, such as a laptop in a hotel, is plug it in and boot it up. If you can actually get useful wireless in a hotel, you don't even need to plug it in.

Things are different for adding a device to a fieldbus network. The device needs a tag name and a bus address. If the device had been configured and was in service at a bus address, it will restart from power loss with all of that information. If the device is a replacement from the spares depot, it may have a factory default configuration which will not do any process control. Connecting it to the bus will cause it to start with a temporary bus address (one of eight available for the purpose) and enter the uninitialized state. No function blocks are executed.

A device goes through three states to become operational:

1. An uninitialized device provides no communication services except to accept a new tag name or respond to a request for its

identification, which is a unique serial number. A host may send the identification query to the new device at its temporary address to determine if this is a known device that can automatically be given a tag name, or if human input is required. When the host has a tag name it writes it to the device, which enters the initialized state.

2. In the initialized state, a host may assign a permanent bus address. This usually comes from a configuration file in the host that is associated with the device tag name. A valid bus address allows the device to go to the operational state.

3. The operational state provides all communication services. If the device is new, the function block application data for the device tag name must be loaded before function blocks can be executed. The blocks will not execute without a link schedule, so that is loaded last. Now the device is ready to do its part of the process control application.

High Speed Ethernet   The H1 fieldbus has a practical maximum of 32 nodes, even though there are 255 addresses. Fluid processes with hazardous atmospheres (refineries) are limited to six devices by the power requirements of intrinsic safety. In practice each bus has only one control valve because the bus is not redundant, plus an additional one or two transmitters. Lasher called this the "chicken foot topology." It is shown at the bottom of Figure 7.8 at the field end of each bus. Each bus terminates in the control room at the I/O rack for the controllers, just as it always did for electrical signaling. The distance between the control room and field is large to assure survival of those in the control room should some large thing in the field explode. The savings in cable when fieldbus can have multiple devices on one cable is considerable.

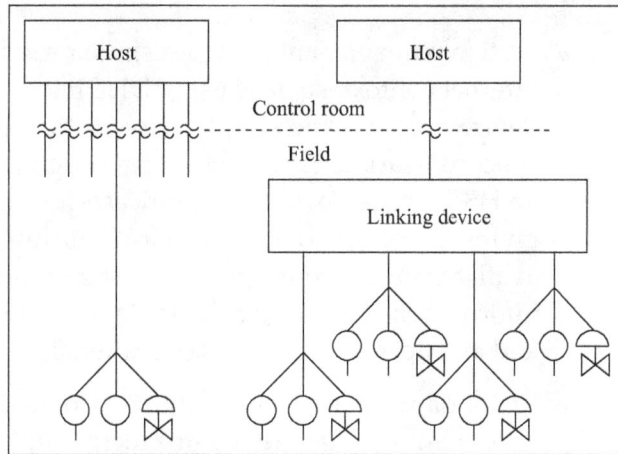

Figure 7.8. Field device topology.

As Ethernet technology advanced to 100 Mbit/s in the late nineties, work began on a high speed analog fieldbus based on high speed Ethernet (HSE), skipping over the H2 phase discussed by SP-50. The first hardware planned was a four-port linking device, shown on the right side of Figure 7.8. Four H1 busses could be plugged into the box, and the data on those busses was

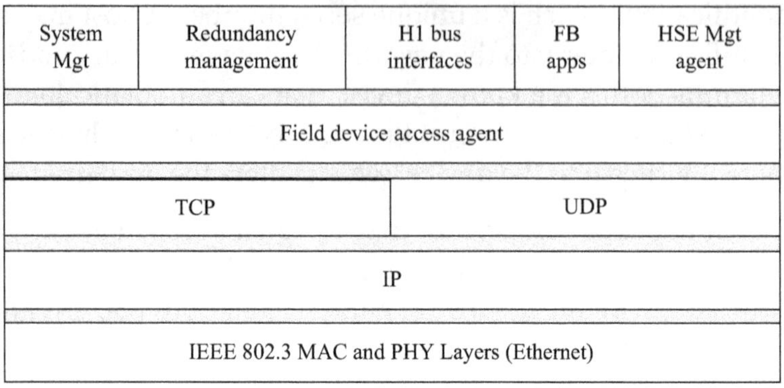

System Mgt	Redundancy management	H1 bus interfaces	FB apps	HSE Mgt agent
Field device access agent				
TCP			UDP	
IP				
IEEE 802.3 MAC and PHY Layers (Ethernet)				

Figure 7.9. HSE protocol stack.

made available on an HSE connection. The box would not be in a hazardous atmosphere so it had no power limitation, which meant that commodity desktop microprocessors could be used. The use of commercial off-the-shelf (COTS) components from memory chips to Ethernet network hardware kept the price competitive.

A simplification of the HSE protocol stack is shown in Figure 7.9. The bottom three layers are standard Ethernet and TCP/IP protocols. The field device access (FDA) agent basically converts H1 FAS layer services to those of TCP and UDP, which is more easily said than done.

The services of the FDA layer are used by the five applications above it, as follows:

- System management provides similar services as in H1 but in an Internet context, such as using DHCP to get an IP address when the device starts up and simple network time protocol (SNTP) for time synchronization. Since DHCP can assign random IP addresses, when an HSE device starts up it multicasts an "annunciation" message giving its device ID and current IP address. No other services are available until a configuration host gives the new device assignment information, including a device tag. As in H1, once it has a tag name and an address, it becomes operational.

- Redundancy management allows two linking devices to function as a redundant pair, in one of three increasingly secure configurations. A linking device that supports redundancy has two Ethernet connections but only one set of H1 connections, to allow either Ethernet bus to fail. A pair of such devices may be connected to the same set of H1 busses to get linking device redundancy. A host may query the secondary device to check on its health while the primary is active. Separate physical Ethernets may be used to maintain the redundant communication interface. The first DCS (1975) sensitized customers to the need for two cables. During a marketing demo at the home office the presenter would dramatically use a fire axe to cut one of the data highway cables to show that the operator maintained visibility of the process. Normal practice would be to separate the two cables as much as possible to prevent one accident from breaking both cables, but there were installations where they were run side-by-side.

- The "H1 Bus Interfaces" application translates H1 message structures to HSE message structures. The sizes of some variables were increased to take advantage of the increased processing power in the linking device. Also, bridging may be provided so that one of the H1 ports may exchange messages directly with another.

- One or more function block application processes may be included at the HSE level, allowing the linking device to function as part of a distributed control system with H1 I/O.

- The HSE Management Agent performs H1 functions and Internet functions such as simple network management protocol (SNMP) and making its error statistics available to other HSE devices.

Linking devices began interoperability testing in 2002, so most of the bugs have been worked out by now. They did not become the new paradigm for distributed control for several reasons. Foundation Fieldbus lost the spotlight to wireless technology, partly because change management at plants using fieldbus didn't always work. There is a jump in complexity going from 4-20 MADC to digital communication, starting with the extra care required in the physical installation of the busses. Large users like refineries had few problems, but smaller companies foundered on such things as existing work rules that didn't allow instrument technicians to do digital configuration.

Optical Fiber Fieldbus   The physical layer for H1 fieldbus includes fiber optics, so users could and did isolate segments of a bus with commercially available optical thick fiber cable. Thin fiber cable wasn't used because it requires special tools and techniques. One company, Fuji Electric, ran the fiber right to the transmitter. This provides excellent immunity from man-made and natural electromagnetic radiation (noise), including lightning strikes on tall distillation columns. It also eliminates any concerns for electrical safety in hazardous atmospheres.

An optical pressure transmitter was available in 1993, and a 16 port optical star coupler was shown in 1995. A complete system of transmitters, optical star couplers, and optic to electric converters using H1 fieldbus was announced in 1997. The major disadvantage was battery life in the field devices, which was only two years with early lithium batteries. A remarkable optical to pneumatic converter for operating control valves had a battery life of 1.5 years. An electrical current to pneumatic converter drawing 20 MADC for 1.5 years would accumulate over 250 amp-hours, which is 100 times the capacity of a size AA lithium battery, or 25 times that of a size C battery.

For whatever reasons, the system did not become popular. Improved transmitters with four year battery life were introduced in 2005, but the system was discontinued in 2010. Perhaps wireless communication proved less expensive than fiber optics, or perhaps, like Fieldbus, there were too many new techniques to learn.

Wireless Communication

Technology marched on, and churned Marconi's wireless into IEEE 802.11 b, g, and n as radio links for Ethernet (802.3). Technology had to go further to be useful in industrial processes. Early work was not promising, because industrial equipment placement could not guarantee line-of-sight reception, and so guaranteed multipath reception—like a telephone connection with poor echo cancellation. Three things had to be available to allow wireless field instruments to be practical:

- Battery life had to improve beyond five years, and solar panels to supplement them had to become low-cost. The batteries must work in all but the most extreme temperature environments. The availability of lower power microprocessors helped to extend battery life, as did long intervals between transmissions.

- Radio transceivers and antennas had to improve to give reliable reception at industrial ranges for a reasonable cost. In the early days, some of us struggled to get wireless home networks to function in wood frame houses.

- Software had to be developed to compensate for multipath reception and loss of communication as a crane went by carrying a large storage tank. Mesh networks appeared circa 2007 that rerouted broken paths automatically. The IP protocols could handle duplicate messages from multipath reception.

Standardization was necessary, and so we had two of them. The dominant manufacturer of fluid process transmitters developed Wireless HART to complement the extensively used open HART protocol for electrical 4-20 MADC signaling. Wireless allowed users that had wired HART devices to extend the benefits to existing non-HART devices with a modification kit or a separate nearby wireless device. The principle benefit was to add devices to the plant asset management system. Wireless HART was focused on the field instruments without opening the network up to the Internet.

Another major vendor backed the ISA standard ISA-100, which was expanding its scope at a rate eerily reminiscent of the SP-50 fieldbus work. A wireless network can be used for many things besides process transmitter information, such as hand-held terminals that allow field operators to see what a central operator can see. This could return the operators from the central control room to the process itself, as discussed in the Operators chapter. If it is opened to the Internet, there are all sorts of applications, such as video cameras, remote monitoring, and employee location that could absorb all of the bandwidth.

The competing standards are now each IEC standards—Wireless HART as IEC 62591 and ISA 100.11a as IEC 62734, after getting the ISA version approved by the American National Standards Institute (ANSI). Wireless HART was designed by vendors looking for a minimal solution. ISA 100.11a was designed by an open committee to accommodate as many ideas as possible. Fortunately for hopes of interoperability, they both agreed that the physical layer would be IEEE standard 802.15.4 in the 2.4 GHz industrial band. Reading about 802.15.4 reveals that it was designed for short range wireless personal area networks, which would be used if wearable computers ever became popular. It is a standard with many options, to the point that interoperability is not assured.

As it turned out, interoperability will not be possible without a gateway device. Dick Caro, co-chair of ISA-100, announced in February 2013 that efforts to merge the standards had failed. There was no doubt that the best ideas of both could be merged into a common standard, but it would not be backwards compatible with existing systems. Once again, advances in technology (affordable radios, mesh networks, and 802.15.4) that could not be made by control vendors caused multiple ways of using the technology to be developed. And once again, the existence of a user base for each vendor made backward compatibility a requirement that could not be compromised. The standards process is just too slow to prevent building the user base. It has been said that the only reason that God could create the Earth in seven days is because he did not have a user base.

The situation could be contrasted with the development of the Internet Protocol, but it would serve no purpose. Vendors will probably build devices that can be configured (or downloaded) to use either protocol, since the physical layer is the same IEEE 802.15.4. Had the compatibility requirement been dropped and a merged standard developed, vendors would then have three protocols to support. Another way to accept the failure to merge is that mergers remove diversity that is necessary to deal with diverse situations. There are many discrete process protocols because there is much more diversity in the needs of discrete processes.

Both standards specify mesh networks (but differently) because that method is most resistant to line-of-sight paths being broken by the activities of people driving large machines. Figure 7.10 shows a typical mesh network with one gateway device. Each of the circular field device nodes can function as a router, passing a message through it in the direction of the gateway or backwards. Each link shown in Figure 7.10 is between two devices within radio range of each other. Solid lines represent normal paths and dotted lines show alternate paths that can be used when a normal path is broken. It is left as an exercise for the reader to verify that one broken link will not stop a message from getting to the gateway. Time will be lost, but the message will get through.

When a message passes through a router, it is said to take a hop, and a hop counter in the message is incremented. The gateway tries to configure the mesh for the minimum number of hops, as shown by the

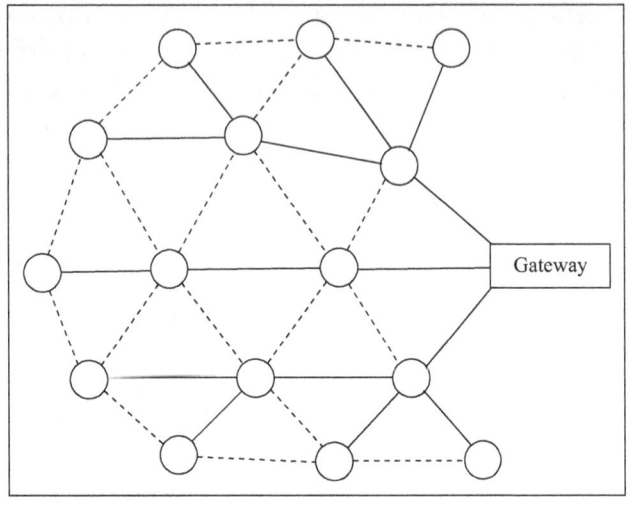

Figure 7.10. Mesh topology.

normal paths. If the nodes were completely free to configure themselves, a message could hop from one field device to another without ever getting to the gateway. If the routers have a maximum number of hops for a message, they will stop forwarding when the hop counter reaches that number, just as they do for the Internet. The IP calls the hop count "time to live."

The Wireless HART System Engineering Guide, Rev 3.0 (publicly available, see References) offers some practical advice from five years of experience with mesh networks:

- A mesh network can offer better than 99% availability if there are at least five wireless devices within range of each other and 25% of them are within range of the wireless gateway to host systems. Each device should be within range of three other devices. If this cannot be obtained in practice, add more devices or repeaters at locations selected to improve the mesh.

- The range varies with the amount of surrounding equipment and the height off of the ground above two meters. A range of thirty meters is typical for process equipment so dense that a truck could not be driven among them. Open that up with vehicle access and the range becomes about 100 meters, and doubles that for a tank farm. Walls can reduce the range to zero. This is for open fluid processes like refineries. If the building is enclosed, such as a power plant or a pharmaceutical factory, then each floor and enclosed room requires a separate network. Placing gateway antennas on the ceiling of an enclosure improves range.

- Network availability will suffer if there are too many devices with frequent updates. This will also reduce battery life. Typically one network with one gateway can handle 100 devices with eight second updates. The maximum number of devices can be reduced by host activity doing remote configuration or frequent polling of devices by asset management applications.

Jonas Berge, in "Wireless Meeting Your Needs" (see References) describes the NAMUR (a group of German users) recommendation NE 124 for wireless systems. It covers availability, interoperability, security, power, integration into existing systems, forward and backward compatibility of device versions, network management, diagnostics, configuration, commissioning, device replacement procedures, and certification.

NE 124 classifies control applications in three levels:

Class A. Functional Safety—time-critical applications required by functional safety standards

Class B. Process Control—time-critical applications requiring high availability and reliability

Class C. Indication and Monitoring—applications that are not time critical.

Nobody recommends class A for wireless, so wired systems are required, and few recommend Class B for wireless given the consequences of leaving a valve in the wrong position. Class C applications can improve the knowledge of a process, and perhaps reduce steam leakage, or unexpected tank level changes, or prevent environmental disasters. These things have value at the bottom line and so may justify deployment without the costs associated with wiring.

Wireless communication is not well suited to closed loop process control, due to the uncertainty of timely delivery of information during a function block execution cycle. As of this writing, there are no wireless analog output devices. Digital output devices are available, but it may take 30 seconds from command to execution. Shades of MAP, and for a similar reason. A mesh is not a ring topology, but messages may have to pass through devices on their way to the gateway, and may have retries and reconfiguration of the mesh that delay the message.

Jonas goes on to describe many factors in choosing a wireless network, including the device description language and network management. A network cannot be managed without diagnostics, such as missed and discarded update messages, path stability, reliability, signal strength, latency, number of automatic reconfigurations, and battery status. An essential feature is a "live list" of devices that have joined the network, with timestamps for appearing and disappearing. Deviations from normal conditions must be reported, hopefully to someone who can do something about them.

Given all of the advantages of wireless over wired, it is only fair to point out that a wired system has none of the possible wireless network management problems, especially not battery status. A wired network will have some of the problems, but control messages will always get through on time.

Security

Diplomats use protocols to negotiate with foreign diplomats, focusing on one country at a time. Often, there are items on the agenda that are hidden from the other side, creating secrets that must be protected. This requires both physical protection for the embassy and its people, and the use of encryption to protect document contents and communications that travel in public.

A manufacturing facility must be physically protected from those who might harm product quality, equipment, or people, or prevent timely delivery of products. Normally, the facility is secured by a fence with a gate that is guarded, to restrict entry to employees. Communications relating to industrial processes that travel in public, either by wireless or the Internet, may need similar security. This applies to mobile devices used for monitoring or control.

Wireless exposes a plant to outside action that bypasses physical security. The standards include a five-digit network name and a 16 octet password that preclude a foreign device joining a mesh without them. There are other ways to make a network useless without the secret codes. Requests to join the network are not encrypted, so the network can be brought down by rapidly repeated requests to join, just as Internet servers can go down under denial-of-service attacks. The radios are low power, so the receivers are sensitive. The receivers can be overloaded by high power broadband 2.4 GHz emissions, which can be generated by an assembly of common microwave oven parts. Beware of the truck with the large antenna parked outside your fence. Finally, it may be possible for an outsider to hear network traffic, isolate traffic from a single device, and endlessly repeat a message containing sensor data. Something like this is done to fake out a wireless security camera, so that the guard sees an unchanging scene while the bad guys swarm over the fence. Some wireless networks allow security to be turned off, which would be irresponsible.

While the secrecy of some data may be an issue, far more important is the ability to send commands to process equipment, especially when improper operation of the equipment could cause danger to the surrounding community. The only people that can send commands to equipment should be those that know how to operate the equipment safely and are aware of what is going on around the equipment.

Process and plant security is far outside the scope of this book. It is mentioned here because there have been cases where the people operating the facility paid no attention to the need for security, thinking that it did not apply to them.

Discrete Process Networks and Protocols

The principle differences between discrete and fluid process networks are their physical extent and their bit rate. Discrete processes with lots of moving parts tend to be protected from the weather by buildings. The cost per square foot of a building assures that there will not be any extra space between process units, leading to more compact networks. Explosion proof housings and conduit are seldom required, so the cabling for a machine may simply be attached to convenient points on the machine. The size advantage may be offset by continuous electromagnetic interference, not just the occasional welder.

While it is common to find control applications distributed across fluid process field devices, this seldom happens in the discrete world. The reason is that the field devices are almost all switches and solenoids, until you get into motion control. If a microprocessor is used for field device communication, it is the least expensive one available, with no room for control calculations. This is because there may be hundreds of discrete field devices in a discrete process for every transmitter field device in a fluid process. More often, the micro is in a device that provides discrete I/O multiplexing for some small number of switches and solenoids.

A programmable logic controller (PLC) (see the Control chapter) is designed to solve Boolean logic equations for all of the discrete sensors and actuators that are connected to it. For example, a brewery for a popular beer may have 100 holding tanks for products in various stages of processing and aging. Each of the tanks has valves that connect it as source or destination to a number of transfer headers, as well as clean-in-place fluid. The noxious cleaning fluid must never be directed to one of the product transfer headers, even though the valve arrangement at each tank makes this possible. Each tank has 30 I/O points, for 3000 total. One PLC solves the Boolean equations that prevent mixing cleaning fluid with beer, and does it 120 times per second. It is not possible to break the problem up into something smaller. Well, maybe each tank could manage its own I/O to prevent that, but it would have been more expensive.

A PLC does more than Boolean logic. It can manipulate numbers in the usual ways and compare them to produce Boolean results, as is done for timers and analog inputs. Thus messages transmitted or received may have numeric data types as well as binary values. The messages do not have to be synchronized because an external message does not carry information that the PLC needs for its next cycle. This is not true for remote I/O. It is also not true for motion control, where the movements in multiple axes must be precisely coordinated with individual motion controllers in order to produce the desired contour when machining a part.

It is not always practical to wire each sensor or actuator all the way back to the PLC, so remote I/O units were developed once suitable microprocessors were available. A terminal strip with a reasonable number of terminals (not hundreds) is connected to a processor that samples the inputs and drives the outputs synchronously with the PLC execution cycle. The connection is made with a high speed serial link to avoid delaying the PLC. High speed means small messages, with no room (or need) for the additional data that goes with a fluid process measurement. The protocol is usually proprietary so that the vendor can guarantee performance.

There are, then, two kinds of discrete network—one that carries mostly I/O traffic (device level), and one that carries more complex messages among controllers (controller level). Communication among PLC devices and host computers is necessarily interoperable and standardized, but there are many standards. See the discussion of many standards in the Wireless Communications section of Fluid Process Networks and Protocols earlier in this chapter. In fact, the Wikipedia article titled "List of Industrial Protocols" listed more than 30 protocols, and didn't have them all. This chapter will discuss a few highlights. It all began with serial communication.

Early Serial Communication

Telegraphy is serial communication, using a single wire with earth return or a pair of wires. Morse code required skilled telegraphers, and so a community of people with the right skills arose. They were challenged in 1874 by a five bit code invented by Émile Baudot, which required no knowledge or skill with Morse code. Instead, an operator pushed a selection of five keys to send a symbol (character). This was still a code that required memorization of the five keys for each character, which became International Telegraph Alphabet No. 1. It didn't take long for machines to be developed that would translate standard typewriter keystrokes to Baudot's code, and translate that code to printed characters. After that, anybody that could type could send a message, and skills with Morse code were no longer required. Five keys were not enough to send lower case letters, so the seven bit American Standard for Information Interchange (ASCII) was introduced in 1963 with a major revision in 1967. ASCII was supplanted by UTF-8, developed from Unicode in 1992 for the many character sets used on the web. UTF-8 is backward compatible with the 127 symbols in ASCII, but expands the range to more than a million symbols.

And so we acquired a new name for a unit of serial communication, the Baud, which represented one symbol no matter what means was used to transmit it. A symbol is more general than an alphanumeric character because it includes such non-printing characters as carriage return, line feed, and bell. When computers began using modems to send information over the public telephone network, whose bandwidth for voice was 300 to 3000 cycles per second (this was long before the name Hertz was adopted as a unit of frequency), the limit was about 2400 Baud. Techniques for modulating the telco bandwidth with multiple carriers and phase modulation finally reached a limit of 56,000 Baud.

Exceeding 56,000 Baud required true digital communication such as Integrated Services Digital Network (ISDN) that went from meaning "I Sure Don't Know" to telco executives to "I Smell Dollars Now" in 1989. This was succeeded by Digital Subscriber Line (DSL), which is still in use, competing with digital cable to the home and less often with optical fiber to the home. The symbols in digital communications are conveyed by binary bits as ASCII or UTF-8. The Baud unit has been replaced by digital bits per second, written as bit/s without the plural "s" or kbit/s for thousand bits per second or Mbit/s for million bits per second. Optical fiber has made it possible to go to Gbit/s representing 10^9 bits per second.

The first general standard for serial digital communication between devices appeared in 1962, as Recommended Standard 232. It was designed to connect a telephone modem to a teletype machine. The standard described connectors, pin assignments, and voltage levels. The standard wasn't stringent because the bit rate was 110 Baud (550 bit/s) for an electromechanical typewriter. Also, the distance was 25 feet or less since the modem was seldom far away from the machine. Speeds increased with time, especially when an integrated circuit was developed for RS-232. Personal computers used RS-232 for serial ports that connected to modems and other devices at speeds up to 19.2 kbit/s (now much higher). This is not an industrial strength communication standard.

RS-422 was introduced about ten years later, which had a maximum length of 1500 meters and a maximum rate of 10 Mbit/s, but not both at the same time. Only 100 kbit/s was allowed for the maximum length. Up to ten devices could operate on one cable, whereas RS-232 only had two devices per cable. The cable used twisted pairs and differential signaling for noise rejection, making it an industrial solution.

RS-485 took advantage of integrated circuit drivers whose transmitters could be disconnected from the line to allow up to 256 devices to be connected to the twisted pair cable. The maximum length was reduced to 1200 meters, with a rule of thumb that the bit rate times the length in meters should not exceed 10^8, or 2 Mbit/s at 50 meters. The maximum rate is 35 Mbit/s at ten meters and the minimum is 100 kbit/s at 1200 meters. This is definitely an industrial standard for a network's physical layer.

All of the above standards are for electrical characteristics, and so are in the ISO physical layer. Entire protocols were built on them for industrial use, seemingly one or more for each PLC vendor, and each designed differently. This is in the process of being sorted out, but each protocol has a significant user base that cannot be moved to the next new thing.

Discrete Manufacturing Protocols

A series of protocols has been invented to satisfy the low data, high frequency requirements of discrete processes. Here are a few that were developed before Ethernet became fast enough to be accepted for process control.

Modbus   The Modicon 184 was arguably the first PLC, developed in 1968, using integrated circuits. At first, it was a self-contained machine for solving Boolean equations written as relay ladder diagrams, a well-established method for designing relay logic. The need for communication with other devices became apparent, and the result was Modbus.

Modbus originally used the RS-422 physical layer to carry a relatively simple master-slave protocol with address space for 247 devices. The master sent a command with a slave address to cause the slave to respond to the request. Slaves never initiated communication, and so there were no traffic collisions. This meant that the master had to poll each slave in turn to see if it had anything to say, even something alarming, which kept the number of slaves on a bus well below 247 for the most part. Address zero was reserved for a broadcast by the master, but there was no provision for slaves to reply other than to a request for identification of a single device.

Modbus messages contain a function code in addition to the slave address, which is divided into three categories: data access, diagnostics, and other. Data access allows read or write to individual bits (contacts or coils), 16 bit registers, or file records. Diagnostics includes getting event logs and identifying devices. 'Other' handles "encapsulated" transport where the data only made sense to an application that knew what the bits meant, but large binary objects are not supported. Messages can be interpreted as binary or ASCII printable data.

As field devices such as chromatographs began using microprocessors, Modbus was the only available protocol. Data calculated by the device was made to fit the Modbus data types of contact, coil, and register. A device did not necessarily talk to a Modicon PLC, and what it said was not necessarily intelligible to devices that were not made by a particular vendor. Any computer with an RS-422 port or RS-232 converter could capture the bytes, so several protocols were developed for field devices.

The protocol developed a large user base because it was free and open, and because it placed few restrictions on device manufacturers. The disadvantage of few restrictions is that a host device has no way to discover the properties of the device or the data, except for the host to be pre-programmed with that information.

CAN Bus   The automotive industry produced the first digital communication method primarily intended for sensors and actuators. Bosch in Germany, founded in 1886, makes electrical auto parts such as ignition systems, and is still in business with a wider product line. Bosch engineers developed the controller area network (CAN) to provide a single serial bus to replace the large and growing wire harness in cars and trucks. It was released in 1985, after the development of microprocessors that could survive in the automotive heat, vibration, and noise environment. CAN became ISO standard 11898 in 1993, and now has dash numbers for improvements in speed and definition of the physical layer. The original speed was 128 kbit/s and is now 1 Mbit/s, as of this writing.

The CAN bus became the basis for several PLC device level protocols, such as DeviceNet.

The CAN bus has elegant simplicity, elegant because of all the work that went into making it simple. Nodes do not have addresses, so each message has a unique message identifier that doubles as the message priority, allowing air bag signals to have priority over raising and lowering a window. There are around 70 nodes in an automobile, most of which control a discrete device such as an interior light or a side mirror adjustment. Further, bus arbitration is handled without a bus master device. All devices have receivers that are always on and transmitters that are only active when the device has something to say. The message identifier is constructed so that the highest priority has an active transmitter for all of its bits. If a second device starts a message it sends its identifier and listens to each bit. If it doesn't hear its own identifier, it immediately goes quiet and waits a predetermined time. The higher priority device does not notice that the second device started and stopped and so its message is sent without error.

All messages contain a space for an acknowledge (ACK) bit at the end of the message. Any device that has correctly received the message asserts an ACK bit in that interval, telling the sender that a device has received the message without requiring the receiver to transmit a separate ACK message. This allows many devices to acknowledge the message without using bandwidth or causing collisions. These bit operations require a bit time that allows the slowest microprocessor to react to each bit on the wire.

There are only four different kinds of message, called frames. A data frame contains data in zero to eight bytes, so a frame can transmit up to 64 binary values, with each bit identified by its position in the data bytes. A "remote transmit request" frame allows a device to query any device that has a message with a specified message identifier. The device that has the message will respond with its data as soon as the bus is free. Such queries are usually used for devices whose priority is so low that response time is not a factor. An error frame is sent by a device that detects one of many possible errors. An overload frame is sent if the receiver is unable to answer because of previous requests or there is so much traffic that the device has no time to interpret the message. Overload frames cause other devices to wait longer before transmitting.

AS-i Bus   The actuator sensor interface bus was developed by a consortium of discrete device vendors in Europe and introduced at the Hannover Messe in 1994. It was designed to provide deterministic communication at high speeds for up to 31 devices with only a few bits of information to transmit or receive.

It is deterministic because each device is allowed a 150 microsecond time slot, so that 31 devices take only five milliseconds to process. Discrete process machinery has many devices which fit these qualifications. A request from the master is 14 bits long. The slave's reply is seven bits long, including four bits of data. A slave has about 20 microseconds to recognize the master's request and begin the reply to fit in the 150 microsecond window. Contrast this with three milliseconds for a fluid process device to send its message.

A special cable is available that has a cross section that constrains an insulation-piercing device to mount with the proper polarity. A cable may be festooned around a machine and the devices attached where they are needed, or moved when the machine is altered, or added when a better device is developed. The maximum length of 100 meters is more than adequate for most machines. The ease of connection greatly reduces installation costs, as well as eliminating the cost of wires from each device to a PLC.

Of course, users wanted more devices, more data, and more distance. The specification was enhanced in 1998 to double the number of devices and double the number of data bits without losing compatibility with the original devices. The extra address bit was taken from one bit of output data, reducing the number of output commands on a bus, and the scan time doubled to ten milliseconds. Even more capacity was added in 2005 as more microprocessor based field devices became available.

The bus has no redundancy. There is only one power supply and one gateway to a high speed network or scanner card in a PLC. This is acceptable for a single machine, because there are many things that may cause a single machine to stop, and the bus is simple and reliable.

As with CAN bus, there is no way for a host to determine anything about the data or the device. The data is simply mapped into the PLC I/O memory in node number order as if it were individual contacts and coils.

PROFIBUS or Profibus   A German consortium led by Siemens developed a Process Field Bus (PROFIBUS) specification that was released in 1989. This was developed during the fieldbus standardization effort that began in 1985. Previous protocols had no or limited ways for a host computer to learn anything about the device or its data. Effectively, one had to have a code book to interpret the bits that appeared in the receiver. The thickest part of the Profibus specification was the application layer, called Fieldbus Message Specification (FMS), which included a dictionary of the communication objects used by the device and the services to read or write that dictionary, if it was writeable. The dictionary contained descriptions of data types and locations for variables that described the device and the structure of any item of data. This concept was accepted by SP-50 as the answer to opaque devices. Well, not all of the eight varieties of SP-50 accepted it. It is part of Foundation Fieldbus as discussed above in the Foundation Fieldbus section of Fluid Process Networks and Protocols.

Profibus DP (decentralized peripherals) is a device level protocol for discrete devices, which has three categories:

DP-V0 is the basic level, as a deterministic, cyclic, master-slave protocol for up to 126 field devices.

DP-V1 adds non-deterministic messages at a lower priority, such as those for reading objects in the object dictionary and writing configuration information.

DP-V2 adds slave to slave communication with token passing at the lowest priority.

Profibus PA (process automation) is based on DP-V1, and adds Application Profiles to attain the controller level. PA became the answer to Foundation Fieldbus when Siemens decided against allowing their users to have control in the field. As things turned out, this may have been a wise decision, given the limited acceptance of control in the field.

CC-Link This protocol was developed by Mitsubishi Electric Corporation in 1997, too late to be listed in the original eight protocols from the ISA SP50 work that became IEC 61158. It offered 10 Mbit/s transmission speed over 1.2 km, extendible to 13.3 km with repeaters. With 64 nodes it could cycle at less than 4 ms, depending on the distance, using a master-slave system with floating master and hot swapping of nodes.

CC-Link became popular in Asia, and standardized in China, Korea, and Taiwan as well as the revised IEC 61158 and 61784. Usage approaches ten million nodes. The CC-Link Partner Association assures interoperability of products, and has testing facilities in the USA, China, Korea, Japan, and Germany.

CC-Link IE is the Ethernet version with a Gigabit transmission rate over optical fiber or standard Cat5e cables. This protocol is useful for both control room and field applications.

DeviceNet and ODVA Another major PLC vendor developed a device level protocol for its products, basing it on CAN and its own ControlNet. It would be just another proprietary network, except that the vendor decided to make its protocols open in the form of the Common Industrial Protocol (CIP) managed by the Open DeviceNet Vendors Association (ODVA), with 274 member companies in 2006. ODVA revises and maintains the specifications and performs interoperability testing to assure users that what they buy will work on a CIP network, in a manner similar to the Fieldbus Foundation for fluid process devices.

CANopen This protocol is typical of second-generation protocols. CAN is a simple message-based protocol with a simple physical layer, as described earlier in this section. CANopen has node addresses and an object dictionary, with standard physical layers. The elegant simplicity of CAN has been replaced with an ISO-layered stack and communication objects, but it is an open protocol. The higher speeds possible with standard physical layers have allowed more complex messages to be handled in the same period of time.

EtherCAT   Ethernet developed a bad reputation for non-determinism in the days before high speed switches. A highly respected process control vendor introduced a new system in 1987 that used Ethernet for controller level communication. They argued that Ethernet was so fast that the result was so close to determinism that communication on a lightly loaded network could be called deterministic. The laughter quickly died as the cost advantages of commercial off-the-shelf (COTS) Ethernet became apparent and proved to work quite well. That started other vendors scrambling to use Ethernet as a physical layer, causing concurrent development of proprietary methods.

There are too many Ethernet-based protocols to discuss here. Most of them are protocols that formerly used RS-485 and now use COTS Ethernet, perhaps with a new object-oriented design approach. One notable exception is EtherCAT. This protocol is designed to copy the contents of a portion of a PLC I/O memory map at a high rate of speed using an application specific integrated circuit that has direct memory access. The device processor is not involved in the transactions. It is a low cost network that can transfer 1000 discrete bits in 30 microseconds with a maximum message size of 12,000 bits.

Wireless and Security

There is less need for wireless in discrete processes than fluid processes because there is less geographic distribution. Current bit rates for wireless make millisecond sampling impractical. The same cost savings for eliminating field wiring apply, except that power wiring instead of batteries is more likely, especially for actuators. Also, the savings will not be as large as those for fluid processes because the wires are shorter and they are not buried in trenches.

AS-i bus and some others provide an excellent compromise between point-to-point wiring to PLC I/O terminals and wireless. The bus can deliver enough power to handle sensors and many actuators.

Wireless is the best choice if the controlled machine is mobile or gets moved to reconfigure for another product. In this case, the machine has a complete control system that can maintain operation without wireless, which is used for higher level command and data messages.

Wireless discrete devices such as emergency stop switches are being offered. Wireless allows them to be installed anywhere, perhaps away from one of the wired networks described above. The same concerns with interference and security apply, which might make a wireless emergency stop look like a bad idea.

Physical security is still essential, as is the security of any Internet connection, and the security of mobile device monitoring and control is still a concern. The need for security does apply to you. Neglect it at your peril.

Gateways

A communications gateway is required to connect one network to another when their networks do not interoperate. The gateway must be a node on each network connected to it, which can mean different media access hardware. If the interconnection must be reliable then redundant gateways are required, with all the problems of making redundancy work. A gateway may have one or more firewalls, but that is not its primary purpose.

One of the first uses for gateways in process control was to connect a brand new distributed control system (DCS) to the superseded control system. The new system would have a new network protocol that made the old network seem primitive. Both the data structures and the protocols were different, so conversion was not simple. Vendors avoided doing this, of course, but control vendors don't have control of computer or network technology. When that technology moves on, the vendors have to move with it.

Another use for gateways is to connect networks designed by different vendors working from different standards, as occurs when one vendor's network must be connected with any other vendor's network in order to have a working control system. The biggest challenge in the control world is to integrate a fluid process network with a discrete network, because the design philosophies are entirely different. The biggest challenge of all is to integrate a control network with a business network, because the people on either side seldom understand what the other is talking about.

OPC UA

Gateways are usually created by vendors or system integrators. There is a need for a general purpose gateway for process control. One vendor set out to fill that need with software, introducing OLE for Process Control (OPC) in 1996, when Microsoft Windows 95 was new, and MS offered Online Linking and Embedding (OLE). The concept of a graphical user interface is taken for granted today, but in 1990 many people were used to using text terminals and command lines. Microsoft Windows with its mouse and menus was a big deal back then, even though the personal computers of the time were so slow that it was called Windoze. The less said about the teething problems of OPC, the better, but there was definitely a need for the product.

Note: What follows is not intended to promote OPC in any way—they don't need it, as it is the industry standard. It is here to illustrate the factors that must be considered in the design of a gateway. Some of the material may have been misinterpreted by the author, because he hasn't taken the extensive training necessary to fully understand OPC. The OPC web site is rich with information.

The OPC Foundation makes specifications, not products. The specifications are developed by committees of vendors and users. The vendors build products that are tested for interoperability using tests developed by the Foundation and modified by what the vendors can actually produce. Initial testing reveals weaknesses in the wording of the specifications as well as misunderstanding by the test developers. This is the same model that the Fieldbus Foundation uses for development of fieldbus specifications, except that OPC can work with any level of device in a process control system and provide connectivity to OPC clients in business systems.

Microsoft replaced OLE with dot NET (.NET) in 2002 as a set of software classes and a hardware-independent runtime environment. .NET was a great improvement over OLE but needed more debugging by customers to be stable. OPC began working on a new version to use .NET, which was released as OPC Unified Architecture (OPC UA) in 2006. The primitive component design of DCOM was replaced by a modern service-oriented architecture. OPC officially became Open Platform Communications in 2011. It isn't just for process control anymore.

The original OPC offers three different ways to access data:

- Data Access (DA) provides read and write commands for live data in control systems

- Historical Data Access (HDA) provides read commands for archived data in a control system's historian

- Alarms and Events (AE) allows a client to subscribe to unsolicited messages from the OPC server when events such as alarms or changes in value or state occur in a control system.

Unified architecture combines these access methods and more into a single set of many services. The services are based on web standards for open communication with a wide variety of platforms and operating systems. UA is not tied to Microsoft, but works with it as well as versions of UNIX and Java on processors derived from Intel or Motorola designs. Further, OPC UA is scalable from stripped down versions for field devices to mobile devices to control systems to MES systems to management systems (ERP, CRM, etc.) that have no problems with XML and its derivatives.

A major improvement in UA comes from working with fluid process field device developers to use the languages and methods being developed to describe devices, so that UA servers and clients can read the description files to fill in their information and data access models. Discrete processes do not have keys to descriptions that can be read from the simple devices, so tag names and other data must come from files used to build the HMI displays.

Other improvements in UA include:

- Configuration of redundant clients and servers.

- Interposing the Internet and firewalls between client and server, allowing access to the cloud.

- Automatic configuration of a server database by connecting it to a control network.

- Use of a standard security model (X509 certificates) so that message security is built into servers and clients.

- Selection of protocols for binary, binary XML, or XML message structures to suit the speed requirements.

- The Windows Communication Foundation by Microsoft ensures interoperability of .Net implementations.

OPC is a large software system, and OPC UA is even larger, but it doesn't approach a recent Microsoft operating system. It does have the problems of program updates and expiring versions.

The following data from the OPC web site are from slides from DevCon (developers conference, here specific to OPC) 2007 where UA was presented. While not current, they give an idea of the size of the system.

Base UA Specifications

Part 1. Concepts—an overview of UA

Part 2. Security—describes the model for maintaining secure connections between UA client and UA server

Part 3. Address Space Model—the building blocks of UA, such as nodes, objects, and events

Part 4. Services—specifies the service methods that UA servers can provide

Part 5. Information Model—the many data objects defined by UA

Part 6. Mappings—specifies the transport mappings and data encodings supported by a UA server

Part 7. Profiles—defines application-specific parameters for implementation and certification

Part 8. Data Access—adds the constructs of OPC DA (the previous generation)

Part 9. Alarms and Conditions—adds states to data objects such as alarms and conditions

Part 10. Programs—specifies the support for access to and control of application programs

Part 11. Historical Access—adds the previous historical data access (HDA) and historical events

Part 12. Discovery—related to common directory service protocols such as LDAP (lightweight directory access protocol).

Overview of Service Sets

- Discovery—find OPC servers and their connection endpoints
- Secure Channel—initiate and terminate secure communication
- Session—initiate and terminate client connection to a server
- Attribute—the basic read/write service for values and historical values
- Subscription—used by a client to subscribe to specific data updates by the server
- Monitored Item—another form of subscription
- View—browse the servers and identify nodes for connection
- Query—browse the servers in order
- Node Management—add and delete UA nodes
- Method—enables calls to specific object methods

Using OPC UA Figure 7.11 shows five levels of a manufacturing enterprise where OPC may be used. There are four networks from the point of view of OPC. The device network is for field or analytical devices that contain device-level OPC servers, shown as boxes 13, 14, and 15. Client 12 might be for a device HMI or be the gateway for device data to one of the many control systems available. The control system likely has its own device networks chosen from the many established networks described in this chapter.

The control network is primarily for the data required to control the process and allow operator interaction with the process controls. Server 11 is the gateway for a conventional control system, which has its own HMI devices and auxiliary processors. Client 9 is used for historians and HMI devices that are not already in the system.

The networks are joined by three gateways consisting of UA Client and Server applications running in the same machine, numbered 4, 7, and 10. Each gateway transforms the data from the lower network to one suitable for the higher network, while allowing selected commands to be passed down.

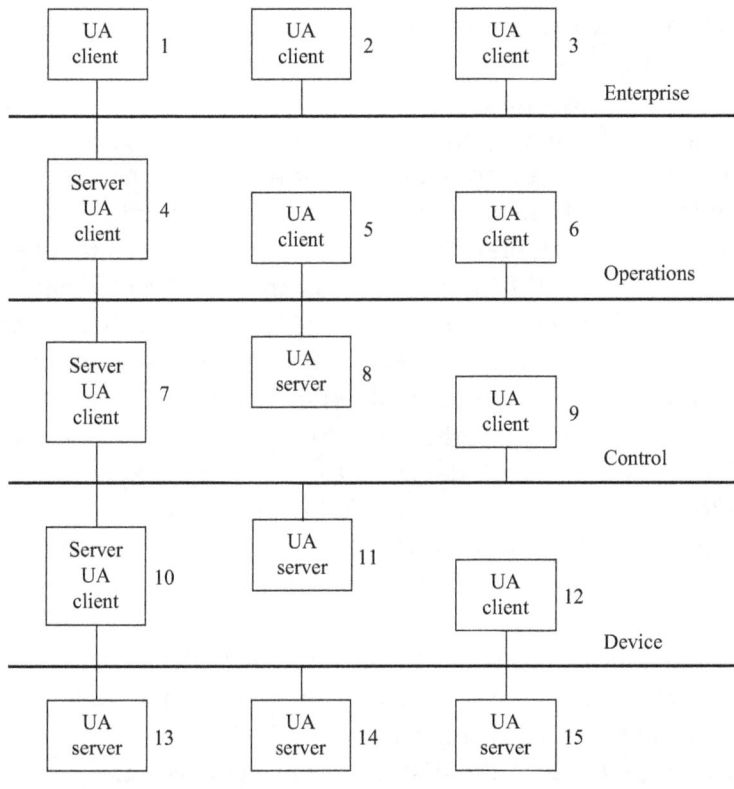

Figure 7.11. OPC topology.

The operations network off-loads the control network with the information technology data required by manufacturing operations management. Server 8 has the data from any conventional control systems. Clients 5 and 6 provide data to the processors that convert control data to operations data, such as key performance indicators and "dashboards." Gateway 4 is a high-level data aggregator, with an aggressive firewall to separate operations from whatever may have gotten into the enterprise network.

The enterprise network uses UA clients such as 1, 2, and 3 to provide manufacturing data to various enterprise functions. These clients may also pass executive decisions to operations, such as production schedules and any business constraints on manufacturing.

This example maximizes the number of UA clients and servers. Your system may be much simpler.

Human–Computer Communication

So far, this chapter has been about communication among machines containing computers. Process control requires interactions with people. The goal for operators is to provide maximum situational awareness in abnormal situations, leading to an effective response. This may require access to the process alarm history, historical process data and the responses of previous operators with outcomes. The goal for managers is overall awareness of process quality and machine efficiencies. Engineers want to know how to get more out of the process. Vendors want to know how they can do better, if they have access to the necessary data. In general, this requires graphic presentations, as many columns of numbers are difficult to accept as input leading to comprehension.

The primary channel for people is vision, and so we have various size displays that have computer-generated images. As discussed in the chapter on Operators, this communication method requires some training and experience to be effective, but it has the highest bandwidth of any of the human senses. People talk to computers with devices like pointers, clickers, and keyboards with much lower bandwidth. It is true that computer speech recognition and generation are becoming quite good, but they are not error free. So far, no control vendor has tried speech, possibly because speech has spawned so many lawsuits, or because it adds a new dimension to the problems of localization. The use of multi-touch screens like those found on tablet devices probably will become popular for changing the size of displays and scrolling, but entries require precision and a confirming click.

The system that drives the display for humans is widely known in the process industries as a human–machine interface (HMI). Good display design is a challenge, for even displays can be misunderstood. A far greater challenge is getting displayable information from all of the different control systems with all of their different communication methods. Fortunately, Tim Berners-Lee designed a system for obtaining data from many disparate computers and called it the World Wide Web in 1989. A key part of that design is the first commercial web browser that appeared in 1993 and the standards that allow a browser client to interact with many different web servers.

Consider that there are many different styles of designing the web pages that are stored on web servers. There are many different computers, operating systems, and display devices and yet most of them can render a web page as the designer intended it to be seen. There is no better method for a manufacturer to provide equipment-specific information to operators, technicians, and others who need that information to solve current problems.

Manufacturers can't provide process-specific information. That must come from displays designed to make the viewer aware of what is going on remotely in the process. Before there were video displays, the information was displayed as a line of process controller boxes mounted in a vertical panel that indicated deviation from control setpoints. An operator could stand in front of the "board" and scan a line of controllers to know within seconds what could be or was causing trouble. On the same panel were trend recorders which served the same purpose with a history of past performances. Sometimes, particularly when there were material transport choices, the controllers and recorders were placed in graphic panel displays built with plastic symbols and pipes that showed material flow in the process. Indicator lamps showed the state of valves and pumps.

This style naturally transferred into computer graphic displays, which were much cheaper to build than physical graphics. But as graphic displays attained higher resolution, some found it hard to resist embellishing the displays with extraneous information such as realistic pipes, brick walls, cut-away views of vessels, and animated fires in boilers. These really spiced up marketing demos for managers, but distracted operators from the simplicity of line diagrams with key bits of information.

Incidents and accidents began to be reported where the operator reported being unable to see key information amidst all of the distractions. Excessive animation has been called "dancing baloney." Users have been looking for guidance from the Abnormal Situation Management (ASM) consortium, Engineering Equipment and Materials Users Association (EEMUA), Health & Safety Executive, and the International Society for Automation (ISA). The ISA finally formed ISA101 in 2006 as a standards committee for better HMI design. The committee bogged down, as committees sometimes do, so a few members wrote a book with the title "The High Performance HMI Handbook" in 2008. Nick Sands of DuPont (if you don't know Nick, look him up and brighten your life) wrote a review of the book with his highest rating and a few caveats.

Perhaps the largest change from existing displays is replacing the black background and bold primary colors with a gray-scale background and subdued colors for anything that is normal. The gray background allows brighter lighting in the control room, which facilitates coordination with other people. Subdued colors allow exceptions to stand out. Also gone are red and green to indicate on and off. Industries differed as to the meaning of red, which is now only an alarm color, so an active device like a pump is now lighter than an inactive device. All of this helps with color blindness.

The next large change is to replace the process graphic pages with an enhanced panel board view of the controlled variables. The enhancements show a vertical bar (because we read left to right—now there's a localization issue) that shows the normal span of the measurement, extended and marked with alarm settings and interlock activation points. If an alarm is active it is shown by a symbol at the top of the bar as both a shape and a color.

Figure 7.12 shows four of the bars. There is a label to identify the measurement at the top of the bar, over an area that is reserved for displaying an alarm condition. The length of the bar is the useful span of the measurement, starting at zero for a flow or the lower interlock trip point. An interlock is designed to stop the measurement from going any further, by closing a valve, or stopping a pump, or starting an emergency shutdown of the process. The normal range of the measurement is shown in light gray, with a triangular pointer. The normal range may go all the way to the bottom. The operator does not have to know any numbers, but the digital value may be shown

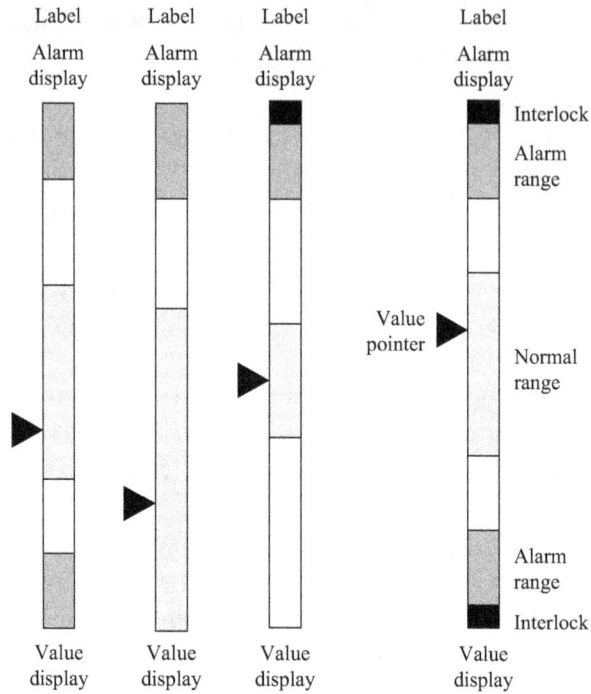

Figure 7.12. High performance bar graph.

at the bottom of the bar. Indicators for discrete control points such as pump motors would be much simpler, perhaps with several shown in the space of one analog bar.

The High Performance HMI has four levels of detail. The first level is the Overview of an entire process, which shows important summary information and performance indicators. It is not possible to control anything from the display, but parts of the process can be selected for display at the second level for investigation and control. The second level is for major parts of the process such as units or cells. The display contains control points, status indicators, switches, and trend displays. A control point must be selected to display a controller faceplate with much the same functionality as a panel board operator had, which is the ability to change control mode and adjust setpoint or output, or to operate a discrete device such as a block valve or pump. An area of the display is reserved for faceplate display, so that no part of the display is concealed when a faceplate is selected.

The third level of display contains the details of an equipment module, such as a heat exchange system or a simple packing machine. The display includes heat and material flows similar to those of existing graphic displays derived from piping and instrumentation drawings. The fourth level may display a single control loop, sensor, or component with space to describe the purpose of the display.

This description of levels merely scratches the surface of what is available in the HMI Handbook. There is a description of a test run for the Electric Power Research Institute to improve operation when the regional power dispatcher calls for an immediate reduction in power because some large load has been turned off. A pair of second level displays were prepared with everything necessary to reduce power quickly, rather than spread out over ten or more traditional graphics. Everybody was happy with the remarkably stress-free operation, where a misstep could ruin very expensive equipment. There are other examples of displays created for handling abnormal situations.

Nick notes that there will be resistance to change, especially to gray scale. As discussed in the Operators chapter of this book, operators expect things to stay put in displays once they have grown used to them. Changing displays is like changing familiar road maps—it's OK to add things, but changing existing roads and symbols will cause trouble. Management may not be willing to take the time and expense for retraining. Nevertheless, the book is full of good ideas for designing operator displays.

References

AS-i Bus, http://www.as-interface.net/

CAN Bus, http://www.canbus.us/

CC-Link, http://www.clpa-europe.com/

Comer, Douglas. *Internetworking with TCP/IP - Principles, Protocols, and Architecture.* Prentice-Hall, 1998.

EtherCAT, http://www.ethercat.org/

Hollifield, Oliver, Nimmo, and Habibi. *The High Performance HMI Handbook.* Plant Automation Services, ISBN-13: 978-0977896912, 2008.

IEC 62591 WirelessHART System Engineering Guide, Rev 3.0, 2012, http://www2.emersonprocess.com/siteadmincenter/PM%20Central%20Web%20Documents/EMR_WirelessHART_SysEngGuide.pdf

IEEE 802 Standards, http://www.ieee802.org/

Modbus, http://www.modbus.org/

NAMUR (not the city in Belgium) http://www.namur.de/index.php?id=21&L=2

ODVA, http://www.odva.org/Home/tabid/53/lng/en-US/Default.aspx

PROFIBUS, http://www.profibus.com/

The Fieldbus Foundation, http://www.fieldbus.org/

The HART Foundation, http://www.hartcomm.org/

The OPC Foundation, http://www.opcfoundation.org/

Programming

Complexity kills. It sucks the life out of developers, it makes products difficult to plan, build and test, it introduces security challenges and it causes end-user and administrator frustration.

—Ray Ozzie (Microsoft, 1995)

Most software today is very much like an Egyptian pyramid with millions of bricks piled on top of each other, with no structural integrity, but just done by brute force and thousands of slaves.

—Alan Kay (Xerox PARC, 2005)

Introduction

Programming for automation is the art and science of telling a machine what to do, so that it will be useful to the people who built it. The Communications chapter described what had to be done to allow machines to exchange information with other machines and humans, using electronic messages. The contents of the messages were of no concern; it was enough to deliver messages to their destination with varying degrees of fidelity. This chapter is concerned with the functions of programmable machines and the contents of the messages that are required to coordinate the functions. It is also concerned with the people who use them.

A piano is a machine for making music. Automation schemes have been created to play pianos, but consider the piano that is inert without a pianist. The human tells the machine what sounds to make through a keyboard with 88 keys and a set of three pedals. The sounds can be quite extraordinary. Classical music requires the pianist to interpret sheet music written (as a program) long ago. Jazz is created on the fly by the pianist, who is now the programmer.

At a high level, many programs are procedures that instruct machines to perform the sequential functions necessary to make a product or perform a major function. These procedures must be open to the users, because operating conditions change over the life of a process. Machines may become available that produce higher quality or quantity, or save significant energy. Users must be able to adopt procedures to use new machines. It is also true that operation of a new process reveals ways to improve procedures. Very few programs or procedures remain identical to their original versions.

This chapter will also address the problem of opaque (as in "black box") automation. Programmers write code that produces predictable results (most of the time), but non-programmers are unable to determine what will happen by reading the code, if the code is even available. The people who operate automated processes are not programmers. When the process behaves in an unexpected way, or new behavior is required to improve the process, they are unable to understand or change the programmer's work.

The fact that a programmer is required means that the user's requirements have to be translated into something that the programmer can translate into programmed machine functions. Opportunities abound for misunderstandings. There is a process known as the V-model that can assure that what the programmer delivers is what the user wanted, as long as the requirements are clear. Requirements that interact in subtle ways will produce results that vary in unexpected ways.

A procedural language that may be used by people who operate processes is necessary. It will require extensive infrastructure programming to make this possible. The infrastructure for open procedures requires an interpreter of the procedure document and communications with all of the devices that will be used by an open procedure.

A control communication language is required that will allow functions created by many vendors to exchange messages. The Foundation Fieldbus model seems useful, as it allows the coordination of control functions in devices made by many vendors. It is not the only model, but it is a place to start.

There is an example of the procedures required to invoke and accomplish a process function in an open and understandable manner at the end of the section on procedures.

The chapter ends with the different kinds of programmers and the languages they use, which are described to help non-programmers get some sense of what it means to be a programmer. Real programmers can skip that part or read it for amusement.

Manufacturing Functions

A manufacturing process transforms materials and energy into desired products or intermediates as well as waste materials and energy. One or more process functions are required to carry out the transformation. Additional auxiliary functions that do not change process materials may assist process functions in their work. Utility functions provide common materials and energy to as many process functions as the utility can handle. Transportation functions move materials from one place to another. It is even possible to automate storage functions.

Each function requires at least one procedure if it is more complex than simple regulation or repetitive sequences, such as those functions available in a basic PLC or DCS. The functions and their procedures are normally reduced to a minimum by directing other functions to do the work. This can get quite complex, to the point that it is impractical to implement in one program. It is necessary to establish a hierarchy for the many kinds of functions required. One good reason is to provide hardware locations that can be given addresses so that communications can be delivered to the right function.

What follows is derived from the work of ISA88 on batch process control, because it is a good place to start, sort of like standing on the shoulders of giants. However, just as there is no standard spoken language for the whole world, different manufacturing organizations have given different names to the concepts to be described. You are free to call these concepts anything you want, but you will find the names given here to be useful when talking to people (or machines) that don't speak your language.

A Hierarchy for Functions

Manufacturing is done in physical locations that have been given many different names as manufacturing evolved, starting with manufactory in the fifteenth century. To simplify the naming in this book, manufacturing is done in a manufacturing facility. The adjective "manufacturing" can be dropped when used in a manufacturing context. A facility must have one or more units of production that do the transformations that are required to make intermediates or products, including packaging. A unit is the only place where material is transformed.

When multiple units are required to make a product (by assembling transformed materials), they are contained in production cells, or just cells. When a facility has multiple cells, it needs a production coordinator function that takes requests for products that come from the business and assigns them to cells or units within the facility. If the business has more than one facility, then it is an enterprise or an empire, depending on who is running it. Empires are characterized by lots of acquisitions driven by an unquenchable desire for growth. Either way, the top level needs a production strategist function that decides which facility or facilities can be used to make the product.

See Figure 8.1 for a diagram of the hierarchy above units. No transformations or assembly of materials take place in these levels, but there are plenty of procedures. The higher in the hierarchy, the less likely it is that procedures can be automated. This is true because today's machines cannot innovate by themselves.

Please don't stumble over words like "unit" and "cell" that have many different meanings. Things will become clearer as we progress with the definition of the hierarchy. This isn't a standard, after all. Standards do exist for parts of this work, but what we have here is an overall picture for all types of manufacturing.

Coordinators or strategists need to know the states of the cells or facilities, when they will be available for new work, whether there are adequate raw and intermediate materials for the work in progress and the new work, and whether there will be a place to put the product when it is finished. The states of a cell include unavailable, perhaps because of maintenance or an equipment failure. The state of a facility may be unavailable if all units are taken down for maintenance at the same time or an earthquake has destroyed the facility. Other facility states are busy and available. In the case of busy or unavailable, the time required to become available must be estimated.

The process unit is basic to a manufacturing facility. No products can be made without at least one process unit in the facility. Units are subdivided for the same reason that an enterprise is subdivided: a single object can become too complex for humans to comprehend. All of the layers above a cell are concerned with coordination of production, because none of them can actually make a physical product. A cell exists to contain the units that are necessary to make a finished product. A unit can make something when it is told what to make, providing the request is within the capabilities of the unit. Units that don't have procedures, even if only for starting and stopping, are not particularly interesting.

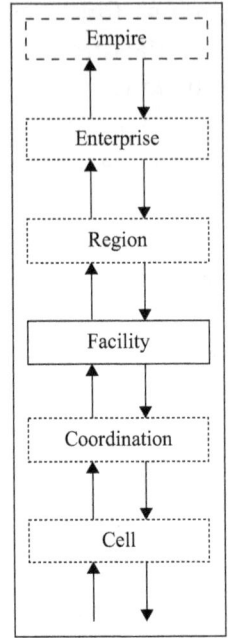

Figure 8.1. Hierarchy above units.

Please see Figure 8.2 for the manufacturing and control hierarchy. No transformations occur without units, and no units can operate without control, if quality is to be assured.

The subdivisions of a unit are called modules. A module is created if it can be reused in other process designs, or if it helps to break down the design into manageable pieces. Again, confusion is possible because there are many things that are called modules or modular. Here, there is always an adjective that goes with module.

A unit often has equipment modules which are collections of

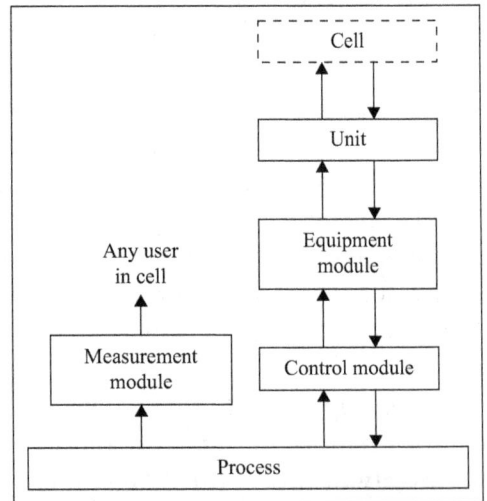

Figure 8.2. Hierarchy below cells.

equipment designed to carry out processing functions required by the unit. An equipment module may have behavior that is determined by one or more procedures, or it may just be a name for a collection of equipment. Either way, the module contains data that is useful to higher levels in the hierarchy. Equipment modules have no direct connection to the physical process, except through control modules.

Control modules can only do control functions. The distinction is that control modules can only perform regulatory control or follow simple fixed sequences to bring a measurement to a setpoint or accomplish a simple process function such as drilling a hole. A control module always has the same behavior regardless of what the unit is doing. Module parameters may be set to modify behavior, such as a setpoint or mode, but there is no selection of procedure names. Control modules may receive settings and commands from equipment modules or units, but the modules are unavailable to any higher level.

Measurement modules simply measure process properties. They may be read by anybody, within bandwidth limitations. No one can command a measurement module to do anything. Modules that only measure, but can be told what to measure, are control modules because they accept commands. Changes to alarm settings are not commands.

It is important that a unit have exclusive access to the equipment and control modules that are attached to it. Otherwise, the unit would have to constantly check to see that no other unit had altered the behavior of one of its modules. A unit may coordinate with other units to perform process functions, generally to allow the transfer of material from one unit to another.

There are other kinds of units besides process units, which have similar connections to their physical equipment and a production coordinator. A utility unit can provide utilities such as electricity, steam, and water to one or more production units, which may coordinate with a utility unit. A common resource unit provides a temporary function to other units. Transportation functions may be common resource units since they connect to two or more units. A resource unit may be required to manage a queue of requests.

Tank storage or warehouse space may be organized into storage units either for the supply or storage of materials or products. Such units may be simple or complex depending on how they keep track of what is stored in them.

Data and Communications

Manufacturing requires communication in the hierarchy, with commands flowing down and status flowing up. Automation with computers requires digital communication, and so each cell, unit, and module must be capable of digital communication. Control modules could communicate with single analog or discrete signals, but that is fading into the past as the advantages of digital communication with a field bus become apparent and cost effective. The advantages of wireless networks are also driving digital communication.

The cell or facility must have a computer that contains application programs that communicate with layers above it and units below it. Note that the layers above it may require human communication. Some of the data concerns the properties of the cell that seldom change, such as its name, communication address, number and names of units, their capacities and capabilities, and other fixed data. This data is called static data, as opposed to the dynamic data of process measurements.

Cells, units and equipment modules have computers with application programs that execute procedures and communicate. Equipment module procedures provide process functionality to the procedures that run in units. The cell or facility may run procedures that request process functions from units in order to make a complete product, or it may simply coordinate requests for services from units.

Manufacturing Procedures

Programming for automation involves creating procedures for moving materials from one place to another in order to physically shape them or chemically change them into new materials. The procedures are written for and executed by computers, but the desired result requires motion and control as opposed to displaying calculated results.

Almost everything we do involves a procedure that can be broken down into smaller procedures. Consider going shopping for a bottle of milk. This involves the use of an automobile to go to a supermarket, find and buy the milk, and return home. This requires knowing how to shop and pay, and understanding names such as "milk" and "Big Box."

Procedure Shop: Drive to destination, locate item, pay for it, and return.

Procedure Drive to destination: Check vehicle, start engine, navigate to destination, park.

Procedure Check Vehicle: Inspect tires, check fluid levels, adjust seat and mirrors, fasten seat belt, doors closed, garage door open.

Procedure Start Engine: Turn on ignition, check indicators and fuel gage level, crank engine, check indicators, handle exceptions such as dead battery, empty tank, and so on.

Procedure Navigate to Destination: Back out of driveway without hitting anything, make the appropriate turns to enter the store parking lot.

Procedure Park: Locate a parking space, park if there are no conflicts with other cars, may require assertive behavior.

Procedure Locate Item: Enter store, find section containing item, locate desired brand of item in desired size.

Procedure Enter Store: Locate entrance and walk through it, acquire a shopping cart, handle exceptions such as store closed or no shopping carts available.

And so on and on . . .

When procedure Shop is used, destination and item must be defined.

Automation Procedures

A unique language called Forth was developed in the sixties for controlling the position of astronomical telescopes. In those days, each telescope positioning system was different, usually because it was done by a different computer, but the need to set a particular position did not vary. Forth allows a programmer to command a telescope to go to a point in the sky defined by a system of universal coordinates regardless of the details of positioning. This is made possible by using layers of procedures that start with the universal coordinates and go on down to the specific commands required by a telescope and its operating system. Go to a different telescope and all you have to do is modify the specific telescope movement routines.

Similarly, every process has equipment procedures that do not vary with the product being made. In the early days, procedures with hundreds of steps would be written to make a product because the procedures directly operated valves and such. It helps to divide the process up into things that can be done by people (or equipment procedures) who know how to operate the equipment. Then they only need to be told the specific requirements for a product.

Consider a set of equipment that can add or remove energy from a stirred reactor. This is normally done with a jacketed reactor but may be done with an external heat exchanger, or both. An operator would need to know the desired temperature of the reactor contents, the actual reactor temperature, the availability of heating or cooling utilities, and the heat transfer coefficient. The operator can manipulate heating and cooling valves, and knows how to transition from heating to cooling and back.

The product procedure tells the operator to heat the reactor contents to some temperature at some rate. Exceptions exist if the reactor is above that temperature, the heating utility is not available, the agitator isn't agitating, or the present heat transfer coefficient will not allow reaching the specified temperature within the specified maximum time. The procedure must deal with these exceptions if they exist. Otherwise, the operator does what needs to be done to bring the reactor contents to the specified temperature.

The act of changing the temperature of the contents of the reactor can be automated by equipment procedures that operate block valves and throttling valves to accomplish the goal, if there are no exceptions. The necessary sensors for reactor contents and jacket temperature must be present. Glass-lined reactors require a maximum temperature difference between contents and jacket that is a function of the lining and not the contents, which is a limit imposed by the equipment.

Hierarchy of Procedures

Hierarchies have proven themselves useful for dividing large problems into manageable pieces. Three levels are enough for procedures:

- Major—the top level is used for procedures that will direct the activities of more detailed procedures that have more specific functions. When a major procedure ends, there is nothing more to be done. A major procedure is similar to the musical score used by a conductor to direct the activities of an orchestra, although the finale may not be so dramatic.

- Minor—the intermediate level is for those procedures that order the use of process equipment procedures but do not accomplish the major objective. Minor procedures are used as necessary to divide the major goal into subordinate activities. There may be any depth of minor levels, but none of them directly operate hardware.

- Equipment—the bottom level contains the procedures that direct the process functions that actually manipulate process materials and energy levels. The goal is to make a process function appear to the higher levels to be independent of the actual equipment used. This allows changes to the equipment without making changes to the procedures above it as long as the process function remains the same.

It may be necessary to refer to a procedure that is above or below the procedure being described. Because there may be several levels of minor procedures, the commanding procedure is called the higher procedure and the commanded procedure is referred to as lower.

The goal for writing procedures is to have as few procedures to maintain as possible. If a facility has a number of similar units, or even identical cells (except for the inevitable differences) then it would be good to have a single procedure that can be run in any compatible unit or cell. Similarly, if there are a number of process functions that use the same procedures with different amounts of materials or energy, then it would be good to have one procedure that can be used for all of them.

In the case of similar units, a major procedure needs additional information to identify the types of units in which it can operate. Equipment transparency is achieved by creating equipment procedures that have a set of interface parameters that match the major or minor procedure directing the equipment.

For example, if the higher procedure needs an energy transfer process function, then a set of interface variables (command and status) are created by control engineers. These defined variables allow equipment procedures to be created that understand those command and status requests. The equipment procedure then causes the equipment to perform the function required by the higher procedure, and to answer its requests for status. If that can't be done, then separate higher procedures will be required.

In the case of similar processes using the same procedure, the major procedure needs additional information to identify the different amounts of materials, energy, and processing times that are required. The equipment procedures are already matched to the higher procedures; only the parameters change.

Procedure Communication

We need to define the messages required between the device that executes a higher procedure and the device that executes a lower procedure if they are in separate devices on a network. This needs to be done in a standard way so that any certified-to-standard vendor could provide the same functionality using the same messages. This requires definitions for nouns and verbs, as well as exception messages. The device then conveys the message to or from the appropriate procedure.

For example, if a major procedure says "Heat reactor <identification> to 42°C at the rate of 2°C per minute," then the equipment procedure must understand all of those words. The major procedure must know the words that will be used to report exceptions, including failure to heat at 2°C per minute.

The use of minor and equipment procedures greatly reduces the number of steps in a major procedure. Further, it brings the procedure closer to a natural language description of the required process functions and their coordination with other functions. This approaches the goal of open procedures, but the goal can't be reached without an infrastructure that allows programmers to write the code that supports the procedures and their communications.

In order to use a lower procedure, the higher procedure must know what capabilities are required from the lower procedure and if they are present. Devices used to sense and manipulate the process must know their capabilities and be able to report them to a procedure. This resembles what personal computers call "plug and play." A printer is asked for its capabilities, and the procedure may modify itself to match those capabilities. This does not imply machine intelligence, but does require standard ways for procedures and devices to negotiate their capabilities.

Processes vary in their need to handle unknown capabilities. A continuous process may have completely defined capabilities which never change, until they are upgraded. A discrete process deals with engineered capabilities that do not change until a better machine comes along. Multipurpose batch processes may be very concerned about equipment capabilities, to the point where the requirements of the procedure may determine the equipment that is chosen for use.

Communication is done by the devices that execute procedures. A message must be addressed to a device name and a procedure name. A calling and connecting protocol is used to assure that a device does not send a command to a device and lower procedure that is not known to the higher procedure. It goes something like this:

A device named Master calls a device named Igor and says, "Igor, this is Doctor Frankenstein. I require your services."

If the Igor called recognizes Doctor Frankenstein as his master, he replies either "Yes, Master" or "I'm busy, Master. Call back in 15 minutes."

If the Igor called does not recognize Doctor Frankenstein as his master, he replies "Wrong number" and sends an alarm. The higher procedure must also send an alarm. If that doesn't happen, the calling device was outside of the system and a security breach has been attempted. Protection against Denial of Service attacks is required by ignoring the caller after some number of wrong numbers.

All kinds of security activities may be attached to the protocol, but they won't stop an insider. They won't stop a wireless listener, if the radio stream is not strongly encrypted. Even encryption may be broken if the attacker is sufficiently determined.

Once the connection is made, the device with the higher procedure passes enough information to the device with the lower procedure so that the task may be done, and requests status reports be sent to the higher procedure on the same connection. When the higher procedure has finished with the lower procedure, the connection may be closed and the lower procedure made available to other callers.

For example, after the connection is made:

Master: "Igor, a storm is coming. Raise the lightning attractor apparatus."

Igor: "Yes, Master." or "I hear and obey, Master."

Igor (reporting status): "Master, the apparatus is in place" or "Master, the apparatus is stuck," which causes the higher procedure to branch to a different procedure that was created for just such an emergency.

Master: "Very good, Igor."

Igor (reporting status): "Master, we have received a direct hit."

Master: "Excellent, Igor."

Igor (reporting status): "Master, it's alive!"

Master: "I'll be right down, Igor. I'm closing the connection now."

Procedures running in units may need to be able to talk to procedures in other units, or even cells, in order to coordinate activities that are usually material transfers. The same calling protocol is required. In this example, the units are Doctor Morbius and Doctor Frankenstein, which are running procedures that require an energy transfer:

> A device named Doctor Frankenstein calls a device named Doctor Morbius and says, "Doctor Morbius, this is Doctor Frankenstein. I require six zillion ergs of energy from you."

> If the Doctor Morbius called recognizes Doctor Frankenstein as his peer, he replies either "Yes, Doctor" or "That exceeds the current Krell capacity, Doctor."

> If the Doctor Morbius called does not recognize Doctor Frankenstein as his peer, he replies "Wrong number" and sends an alarm, just as was done above for Doctor Frankenstein and Igor.

You see how easy this communication stuff is. You should have no trouble adapting this to your own mundane situations. The tricky part is getting the procedure writers to sit down together and agree on a common vocabulary for the words and symbols that will be used for communication.

Other connections are required to allow procedures to use manufacturing resources such as inventory control. The connections are always queries from a procedure to a resource. A resource cannot query a procedure because the procedure may not be available. Resources query the devices that execute procedures, to read status data that has been written by the procedure when it was executing. Commands are not allowed in either direction. A group connection may be used to send status or alarms to a group of human interface devices.

Any device that runs procedures must always have a connection available for adding, editing, and deleting procedures, as long as humans are required to do those things. Security must be excellent.

Creating Procedures

Some of what follows may look like it is attempting to set a standard. Nothing could be further from the truth. Been there, done that, don't want to do it again, but don't regret the friends made or the knowledge gained. What follows are merely guidelines, as some Caribbean pirate said in a movie not too long ago. If you don't like the names for keywords, feel free to use your own, but be prepared to explain them.

A procedure can be written by anyone who knows what needs to be done and knows the process capabilities that are available to do it. A programmer is required to translate that written procedure into something that a computer can do. This book attempts to define procedures in such a way that only one programming effort is required to create the infrastructure necessary to execute procedures written by users. This should allow users to have full understanding and control of their manufacturing procedures.

An example of such an infrastructure is shown in Figure 8.3. The central block is the procedure reader and interpreter. If the computer containing the interpreter can handle multiple program threads, then several procedures may be executed at the same time. They may be copies of the same procedure, so they are referred to by slot numbers or some such scheme.

Communication with other procedures may be required, so there is an incoming and an outgoing message processor. Each processor may need to do language translation when that is unavoidable. Meanings can be lost in translation. An example is an equipment procedure that talks to a manufactured equipment skid that has a PLC that does not speak the procedure language. A start command from the procedure has to be translated into a message to the PLC that sets the start bit somewhere in its program. Users must be able to configure simple translation tables. Systems that use XML and all of its support functions are the direct opposite of simple.

The message processors also have direct connections to five other infrastructure functions. An incoming message may directly set parameter values in tables of parameters for the procedure slots and possibly parameters for the infrastructure. An incoming message may also be an HMI command from a human, which needs to be processed before being passed to a procedure slot.

The block labeled Static Data contains data about the infrastructure that has been configured by a user engineering activity, such as a cross-reference table

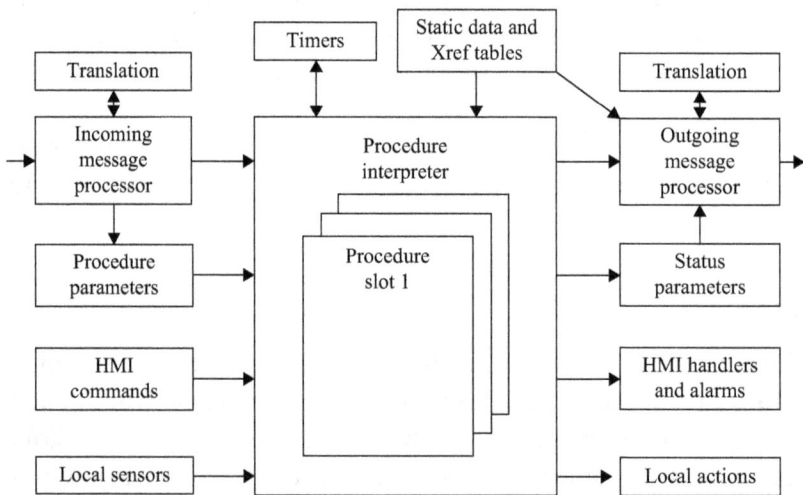

Figure 8.3. Procedure infrastructure.

between generic equipment names in the procedure and the actual equipment that is available to this infrastructure. It also contains data about the capabilities of the infrastructure, such as capacities of tanks and names of procedures that can be executed. The Status Parameters block also contains dynamic data from the execution of procedures and the changing status of the infrastructure. The HMI Handlers and Alarms block contains data required to respond to an HMI query other than status and causes alarm and event messages from the procedure to be sent to appropriate destinations through the outgoing message processor.

If the infrastructure has local connections to process I/O, those messages are sent through the Local Sensors and Actions blocks. All procedures need to be aware of the passage of time, so there may be a separate Timers block where timers set by a procedure are decremented according to social time and notify the procedure when they time out.

Not shown in Figure 8.3 is the communication processor required to handle device configuration activity. This activity is required to make the infrastructure unique on a network and to define its capabilities and options. It is not used during normal operation.

This infrastructure may be included in any cell, unit, or equipment module, as shown in Figure 8.4. If a cell exists, it is directed by the business functions to do something and to report status to the business. The cell organizes units as required to satisfy the request. If there is no need for a cell then those activities are handled by a unit. The unit executes a major procedure to transform materials or to change the state of a process, which directs the activities of equipment modules owned or temporarily acquired by the unit. The equipment modules follow equipment procedures to direct the activity of control modules. Control modules have no named procedures so they have no procedure infrastructure. The only thing lower in the control hierarchy than a control module is the process, which responds to the actions of the control modules.

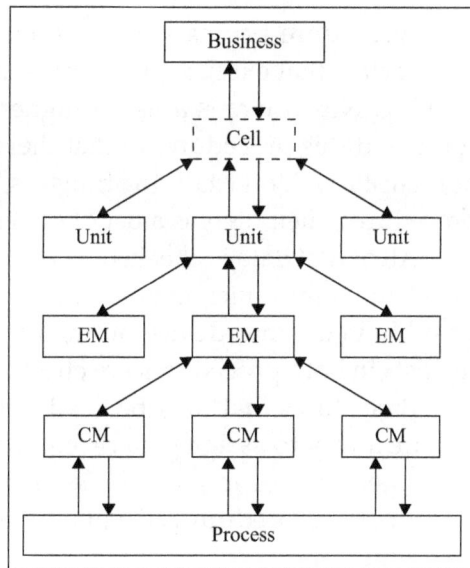

Figure 8.4. Distributed instances of infrastructure.

Procedure Language

Procedures need to exchange information with other procedures, as shown in the Igor examples. The exchange requires that each procedure have a vocabulary of words or symbols that mean the same thing to all procedures.

The procedure authors have to agree on a vocabulary so that the procedures can communicate and behave appropriately. This is part of a larger design effort that selects the process equipment required to manufacture a given set of products. The functions of each module of equipment provide the basis for building a process vocabulary. Any vocabulary requires a dictionary to specify the spelling or symbol and its exact meaning. Machines are not good at double entendre, so a word or symbol must not have a different meaning in another context.

The vocabulary must also contain all of the words and symbols that can be used to write procedures, not just those used for communication. People who write and maintain procedures must make the procedure vocabulary part of their own, with the aid of the procedure dictionary. Note that there is no need to standardize the vocabulary, unless procedures are global within an enterprise. Using the local language will make acquiring the vocabulary less painful.

Words are not the only thing required; there must also be rules for constructing sentences. It may not be possible to develop one set of rules to fit all procedures. This book evades that problem by presenting an example and letting you develop rules to fit your application. The infrastructure programmers would much rather not have to deal with many variations.

Procedure Format

Each procedure begins with its unique name and version identification, so that devices that execute procedures can be absolutely certain of what they are running. Next there is a list of higher procedure names that are designed to command this procedure, so that there is never a mismatch. The primary rule of security is "don't talk to strangers." If this procedure can command other procedures, then there is a list of lower procedure names.

All of that data is checked by the device that executes a procedure in order to set up communications as they are required. When a higher procedure, possibly in another device on the network, wants to use this procedure then the lists in both procedures are checked for matches.

Procedures that may be used with more than one set of equipment will require a way to assign generic equipment names in the procedure to specific equipment names in the process. This is usually handled by the device that executes the procedure. A list of parameters exchanged in communication may be useful.

Non-trivial procedures have more than one state governing their activity in the process. The interpreter can be set to one of a selection of execution states. Some common states are:

> Idle—the equipment is checked for abnormal conditions and readiness to start.

> Starting—the equipment is preparing to run by synchronizing drive positions, heating glue pots, etc.

Suspending—ceasing operation, but ready to go as soon as a condition such as a feed jam is cleared.

Running—normal operation with no exceptions detected.

Holding—an exception has occurred that the procedure can't fix, so it de-energizes the process and asks for human help.

Stopping—a condition has been detected that makes further progress useless, or a stop command has been received.

Aborting—a condition or abort command makes it necessary to de-energize the process as quickly as possible.

Done—the procedure cannot return to Idle without human confirmation that it is ready to do so.

Procedures may have modes of execution to determine how external influences interact with the procedure. Some common modes are:

Automatic—the procedure determines the next thing to do. Human intervention is not welcome.

Single Section—the procedure runs to the end of the current section and stops until a human starts a selected section.

Manual—the interpreter executes one line and stops until a human request starts the next line. When the mode is returned to automatic, the procedure must determine where to resume or the human may be asked to specify a section.

Test—the interpreter runs the procedure, but all messages sent by the procedure are printed to a file, and all incoming messages are taken from a list built to test the procedure. This is also an excellent test of the interpreter.

Simulate—no actual commands are sent to process-connected devices. They are redirected to a simulation application to allow operator training or procedure debugging.

An execution state can be changed by the procedure itself or by a command from a procedure or human, depending on the execution mode. A procedure cannot change an execution mode.

The elements of a written procedure are:

1. Header with all necessary identification and versioning to make this procedure unique.

2. Lists of capabilities and commands, equipment requirements, level of vocabulary, and other static data.

3. List of external data required and locations for the data.

4. Things to be checked and done when this procedure runs for the first time.

5. Things to be checked and done when this procedure's mode is changed from manual to automatic.

6. The body of the active procedure, possibly subdivided into sections that end with "End section." Each section may be subdivided by any of the following headings:

    Check—assure that the pre-existing conditions required for this state are complete.

    Setup—set timers and counters, issue commands to get the required process configuration.

    Start—send commands to start activities and get confirmation.

    Run—assure that operation is nominal while detecting the end of the action.

    Stop—turn off the things that were started and return the process to a neutral state, if required.

    Report—set status variables and send reports to historians, etc.

7. Exception handling sub procedures.

    Assure <test>—sets the execution state to hold if the test fails.

    On <state>—the execution state has changed to the named state. The interpreter jumps to this line.

    Enable Trap <alarm or event>—sets the execution state to hold if the named alarm or event occurs.

    Disable Trap <alarm or event>—turns off the trap.

Note that the hold state is the usual response to an event or condition that the procedure cannot handle on its own. An operator's attention is required. This almost always involves stopping the addition of material or energy to the process. An operator returns control to the procedure by pressing or selecting a Resume button, if it is possible to resume.

Some punctuation is useful in the procedure. The interpreter executes line by line, so the end of a line is marked by a period. Doing more than one thing on a line can cause confusion for the reader (or the error message), so commas and semicolons are not used. Quotation, exclamation, and question marks are not interpreted. The colon is used to start a list of lines, as explained in the next section. The list is terminated by an explicit end line.

Comments are always useful. Comment lines begin with two forward slashes and a space, as shown in the example procedures.

Procedure Execution

A procedure is written in a narrative fashion—one thing follows another. The story begins at the beginning and proceeds until it comes to the end. There are times when the flow of events must stop and wait for something to happen, such as a reply to a message or a time to elapse. The procedure interpreter will stop work on the procedure for a second or so, and then start at that same line to see if the wait condition has been satisfied. If it has, execution proceeds to the next delay, otherwise it waits another second. A procedure interpreter can time-share with other procedures in other slots while it is waiting.

Sometimes the flow of events must proceed along parallel paths that provide separate manufacturing functions that can proceed independently. The procedure creates separate sub-paths that are executed as if they were separate procedures. The line "New path." tells the interpreter to create a new path and to expect a section of procedure that starts with the line "Begin path." and stops with "End path." The interpreter creates the path and then reads the section into the lines to be executed in the path. If the first line is "New path" then it creates another new path in the same way it created this path. A path exists until End path is reached. Each path receives an independent execution cycle until is has to wait for something.

Sometimes a path just ends, but more often the separate functions must be completed before the main procedure can resume. The line "Wait for paths complete." will do that if nothing needs to be done while waiting. If something does need to be done, the line "While paths not complete:" followed by lines of instructions, followed by "End while." will execute instructions that heat, cool, time, or whatever.

Sometimes the flow of events reaches a decision point where a choice must be made. A single decision uses the form "If <expression> then <action>." If the expression is true then the action will be taken, otherwise nothing happens. More choices can be tested by a series of lines of the following form:

"If <expression> then <action>."

"Else if <expression> then <action>."

"Else if <expression> then <action>."

"Else <action>." -OR- "End if."

There may be as many "Else if" lines as required, but the last line must be "Else <action>." or "End if." to avoid confusing the interpreter. After the action is executed for the first true expression, execution will resume at the line following "Else <action>." or "End if."

The flow of events can also be altered by exceptions. The word "Assure" at the beginning of a line followed by an expression causes the test to be made. If it passes, the interpreter moves on. If it fails, the procedure has reached a point where it needs human help, so the execution state is changed to hold and appropriate notifications are initiated.

It is not necessary to specifically request that a message that has been sent has been acknowledged by the recipient because there is an implicit "assure" in the message service. The procedure can also set a watchdog timer that has an implicit "assure" so that the procedure never waits forever for something to happen. Nothing should be left to chance, because the personification of chance is Murphy, who seldom brings things to an agreeable conclusion.

The standard process alarm system can be used to detect exceptions. The phrase "Enable trap" followed by a precise description of the alarm source as to process tag and alarm condition causes the procedure interpreter to examine alarm or event messages on the network for a match. If the trap is enabled by a line in the procedure that explicitly says something like "Enable trap FIC-105 high deviation alarm." then that alarm will cause the procedure interpreter to change the execution mode to hold, no matter where it is in that procedure. The procedure, of course, may set the alarm limits before it enables the trap. The procedure may disable the trap, or let it be disabled by the interpreter when the procedure is done. There is a limit to the rate of alarms that can be tested without overloading the processor. The alarms should be filtered by a reasonable list of group addresses for sources.

It is possible for external programs or humans to change the execution state of a procedure. The procedure should handle the state change with a section that begins with the line "On <state>." When the state is changed, the interpreter will abandon what it is doing, but it will record where it was, and jump to the appropriate line. This is not a "While" statement where the following lines repeat, but a one time jump on a state change. It may be useful to be able to enable and disable the state traps, to avoid jumping out of a necessary sequence of lines. Execution mode changes cannot be handled by the procedure.

Recall that this is not a proposed standard, but merely a set of guidelines for people seeking ways to automate processes. Certainly there are conditions where things should be handled differently. It is also certain that not all conditions have been addressed here. Nevertheless, we press on with an example.

Example Equipment Module and Procedure

Figure 8.5 shows a piping and instrument drawing for a scheme to produce warm water of a desired temperature from sources of hot and cold water. The solid lines represent pipes and the circles represent instruments, where T is temperature, F is flow, and V is a valve. The domed hat on a valve shows that the valve position is variable, while a square hat shows that the valve has only two positions: open or closed. The two-position valve is called a block valve, which is used to change the flow configuration of a fluid process.

Figure 8.6 shows the borders of equipment module Tempered Water 1 and the three control modules within it. This module requires a copy of equipment procedure Robby to function as an equipment module.

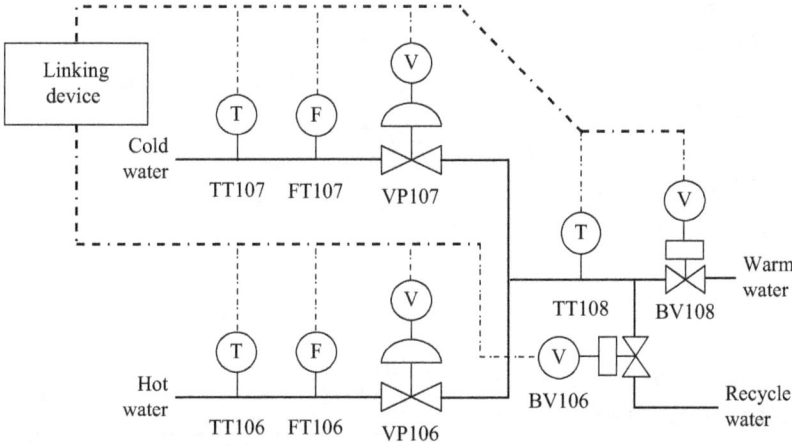

Figure 8.5. Tempered water P&ID.

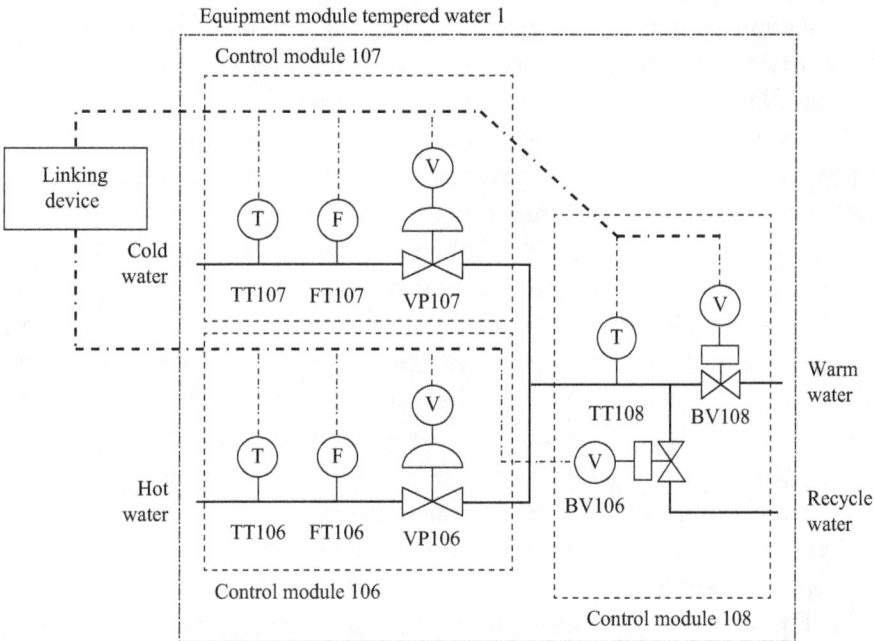

Figure 8.6. Tempered water modules.

The tag numbers identify temperature transmitters, flow transmitters, valve positioners, and block valves. The numbers 106, 107, and 108 identify control groups that are randomly numbered by an instrument layout designer. More elaborate numbers may be used for locations in a process that is organized in a physical hierarchy, such as area 7, cell 22, unit 4, equipment module 2, and control module 5 as A7C22U4E2C5. This is satisfying to hierarchialists but most operators only use the last five to seven characters, in the context of the unit selected by the operator. Some networks may defy the security of hierarchies and use one flat layer to address everything. Such networks require unique complex addresses out to the limits of the network.

The control scheme is shown as a function block diagram in Figure 8.7. There are two control loops, each with a setpoint (SP) that must be set by the higher procedure. The first loop controls the flow of warm water with FC107 by manipulating the cold water flow with VP107 from the sum of the cold and hot flows. The total flow is not measured because the transmitter would be too close to the valves. The second loop controls the warm water temperature with TC108 cascaded to hot water flow controller FC106. The first loop

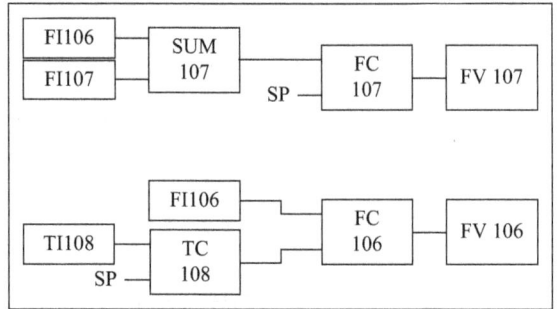

Figure 8.7. Tempered water control scheme.

reduces cold water flow as more hot water is added. Not shown is a feed-forward scheme that uses changes in the incoming flow temperatures to adjust the flows in order to minimize changes to the outlet temperature. There wasn't enough money in the budget to complete the work.

This example adds two block valves to make the equipment procedure example more interesting. The process requires warm water precisely at the desired temperature. This requires a series of adjustments to be made while the warm water is being recycled. When the temperature at TT108 is correct, BV108 may be opened to the process and BV106 closed.

In this example, all of the calculations and PID control are done in Foundation Fieldbus certified field devices, available from many instrument vendors. You have other choices. The dashed lines in Figure 8.5 represent fieldbus wiring, where wires to devices are dropped from a fieldbus segment, shown in heavy dashed lines. If the control loop is important to the process, there is usually only one valve on the segment. An operator can manually control one valve in an emergency. Today's stripped-down plants would not have a second operator to control a second valve.

The Foundation Fieldbus segments end in a linking device, which acts as a gateway from two to four fieldbus segments to an Ethernet connection to a high speed switch. Ever since the days of redundant control highways in the first DCS, people feel more comfortable with redundant Ethernet cables, even if they run them in the same conduit. Most linking devices have redundant Ethernet connections. It is also possible to use two linking devices to provide fieldbus segment redundancy. Very few people use redundant instruments, however.

The Ethernet connection with the linking device is not shown in Figures 8.5 or 8.6 for lack of space. Figure 8.8 shows the linking device with Ethernet connections to other devices and procedure names within the linking device and computing resource. The linking device actually connects to two similar warm water utilities, with equipment procedure names Robby 1 and Robby 2 as two instances (not different versions) of procedure Robby.

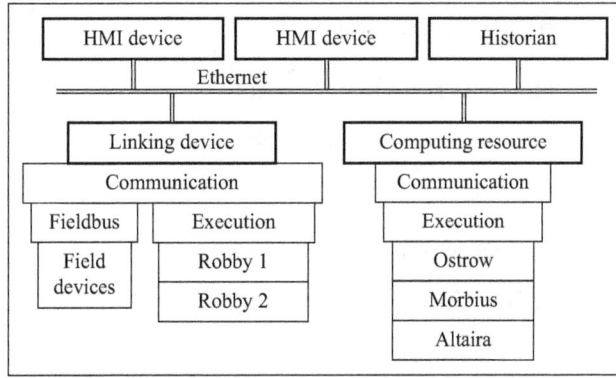

HMI device	HMI device	Historian
Ethernet		

Linking device		Computing resource	
Communication		Communication	
Fieldbus	Execution	Execution	
Field devices	Robby 1	Ostrow	
	Robby 2	Morbius	
		Altaira	

Figure 8.8. Linking device network connections.

The Ethernet side of a linking device connects it to a network of higher-level control devices, consisting of special-purpose computing resources such as operator consoles, advanced controllers, and historians. In terms of executing procedures, equipment procedures could be interpreted in linking devices, although there is no demand for this yet. Higher procedures could be interpreted in advanced controllers or any suitable computing resource.

It is true that equipment procedures could be interpreted with higher procedures in some higher-level computing resource, but this leads to single points of failure. A design philosophy that puts equipment procedures in highly reliable linking devices (they have to be as reliable as the field devices and they are simple enough to be fully testable) allows a higher procedure to start an equipment procedure and leave it alone until it is finished.

With this separation, if the higher procedure or the device executing it suffers a fatal accident, the equipment procedures will run to a passive conclusion and await further instructions. The higher procedure never monitors the flow or level or temperature in order to tell the equipment procedure to close a valve. The equipment procedure is given a goal and told to report back when it has been achieved.

The higher procedure does run a timer in case the lower procedure or the device executing it has a fatal accident. At that point, the higher procedure cannot proceed because at least one of the resources that it requires is not available. Human intervention is required. The higher procedure has to know how to recover from human intervention, as does a lower procedure.

Partial Higher Procedure

This fragment of a higher procedure interacts with the equipment procedure that follows it. The higher procedure manufactures a part called the workpiece, then washes, dries, and paints it. The procedure is not optimal because it doesn't use multiple paths to stabilize the warm water temperature just in time for the workpiece to arrive. This procedure is not the heart of the example, so there is no point to making it more detailed and complex.

Procedure: Morbius.

Higher procedures: None.

Lower procedures: Robby.

. . .

Section is Wash.

Subsection is Setup.

Assure presence of soap.

Assure presence of brush.

Assure presence of workpiece.

Set watchdog to 5 minutes.

Acquire Robby.

Subsection is Start.

Set watchdog to 15 minutes.

Call Robby, set flow to 20 liters/min, set temperature to 52°C, set state to Starting.

Wait for Robby to confirm startup complete.

Subsection is Run.

Set watchdog to 15 minutes.

While workpiece is not clean.

Call Robby, set state to Running.

Wait for Robby to confirm Running.

Insert workpiece.

Wash for 2 minutes.

Rinse for 2 minutes.

Call Robby, set state to Suspending.

Wait for Robby to confirm Suspended.

Dry workpiece.

Inspect workpiece, set clean or dirty.

End while.

Subsection is Stop.

Set watchdog to 5 minutes.

Call Robby, set state to Stopping.

Wait for Robby to confirm Stopped.

Release Robby.

Remove workpiece.

Clear watchdog.

End section Wash.

Section is Paint.

. . . .

Equipment Procedure

This procedure animates the class of equipment modules called Tempered Water. The process is measured and manipulated through control function blocks in Fieldbus Foundation field devices. The higher procedure manipulates this procedure by changing its execution state with messages that can also set parameters. This procedure sends a message to the higher procedure when the requested action is complete.

When the procedure is loaded, it runs through the lines in Initialization and enters the Idle state. The higher procedure changes the state to Starting, then Running, then Suspending, and finally Stopping. Exception handling is done with Assure statements. Watchdog statements are used so that this procedure does not allow cobwebs to form on the machine while it is waiting for something that will never happen.

Procedure: Robby.

Higher procedures: Morbius, Altaira.

Lower procedures: None.

Equipment Tags: BV106, FT106, TT106, VP106, FT107, TT107, VP107, BV108, TT108.

Function Block Tags: FI106, FI107, SUM107, FC107, FV107, TI108, TC108, FC106, FV106, BVC106, BVC108.

. . .

Initialization

　　Set FC106 to manual and 0% output.

　　Set FC107 to manual and 0% output.

　　Close BVC106.

　　Close BVC108.

　　Set state to Idle.

End initialization.

Begin procedure Robby.

While state is Idle:

　　Assure that FC106 is manual and 0% output.

　　Assure that FC107 is manual and 0% output.

　　Assure that FI106 is less than 1%.

　　Assure that FI107 is less than 1%.

　　Assure that BVC106 is closed.

　　Assure that BVC108 is closed.

　　End while.

On Startup.

　　// First, warm up the equipment to get steady temperatures.

　　Set watchdog to 5 minutes.

　　Open BVC106.

　　New path.

　　　　Begin path.

　　　　　　Set FC107 to manual and 20% output.

　　　　　　Wait for FI107 greater than 20%.

　　　　　　While TI107 stabilizes:

　　　　　　　　Assure that FI107 is above 15%.

　　　　　　　　End while.

　　　　End path.

　　Set FC106 to manual and 20% output.

　　Wait for FI106 greater than 20%.

While TI106 stabilizes:

Assure that FI106 is above 15%.

End while.

Wait for paths complete.

Set watchdog to 7 minutes.

Calculate the value expected for TI108.

Assure that TI108 is within 1% of the calculated value.

Calculate the flow setpoints for FC106 and FC107 to give the required total flow at the required temperature.

Set FC106 to calculated setpoint and auto.

Set FC107 to calculated setpoint and auto.

Wait for FC106 deviation less than 1%.

Wait for FC107 deviation less than 1%.

While TI108 stabilizes:

Assure that FC106 deviation is within 1%.

Assure that FC107 deviation is within 1%.

End while.

Set FC106 mode to cascade.

Wait for FC106 deviation less than 1%.

While TI108 stabilizes:

Assure that FC107 deviation is within 1%.

End while.

Call higher procedure, confirm startup complete.

Set state to Suspending.

On Suspending.

Set watchdog to 15 minutes.

Open BVC106.

Close BVC108.

Call higher procedure, confirm Suspended.

While state is Suspending:

Assure that TC108 deviation is within 1%.

End while.

On Running.

    Open BVC108.

    Close BVC106.

    Call higher procedure, confirm running.

    While state is running:

        Assure that TC108 deviation is within 1%.

        Assure that FC107 deviation is within 1%.

        Assure that FC106 deviation is within 1%.

    End while.

On Stopping.

    Set watchdog to 2 minutes.

    Open BVC106.

    Close BVC108.

    Set FC106 to manual and 0% output.

    Set FC107 to manual and 0% output.

    Wait for SUM107 to equal zero within 1%.

    Close BVC106.

    Call higher procedure, confirm stopped.

    Set state to Idle.

    End procedure Robby.

Programs

Computers seem to be able to do marvelous things, but in fact they are merely pieces of hardware. A computer is designed to read instructions sequentially from memory and execute them. A set of instructions is called a program, which is a specific kind of procedure. A computer is inert without a program to execute. The people who construct programs for computers to execute are called programmers and what they do is called programming. The following describes how a computer comes to life, so to speak, as an example of a basic use of programs.

    Any digital computer has a register that contains the address in memory of the next instruction to execute. When power is applied to most microprocessors, the instruction address register is preset to a standard address that is common to all microprocessors made by a company. The address is in a read-only memory chip called the BIOS (basic input output system), and it is the first address of a program that performs the POST (power on self test). The next programs add drivers for devices that are not in the BIOS and then load the operating system from standard locations on different types of rotating memory.

Once loaded, the operating system takes over and indicates to the user that it is ready to accept a command or start an application program such as a word processor. The user then assumes what control over the computer the programs stored on disk have made possible. A programmer may use the computer to create new programs. A non-programmer is at the mercy of the programmer(s) that wrote the chosen application. Success is sometimes measured by the degree to which the user can think like a programmer.

What makes the programs interesting is that conditional branching is possible, as well as the mathematical operators, so a computer can make decisions based on its calculations without having to present calculations to people who will make decisions. Further, a computer can use communications networks to exchange information with other computers and to control peripheral devices for input, display, storage, and output. Some of those peripherals may aim weapons or control industrial processes.

Computers have not taken over the world because they cannot write their own procedures, yet. Humans have to think of things for computers to do, and then design programs to make them do those things. There are many kinds of programmers from specialists to generalists. There are many programming languages, some of which are faded fads while others have proven generally useful, such as C, C++, Java, Perl, and Python.

Creating Programs

The need for a program begins with a requirement for some task that can be done by a computer. First, the requirement needs to be defined so that it represents the goals of the user. Otherwise, a programmer can become like one of those people who think their job is to fight alligators when the goal was to drain the swamp. While a single programmer may write a simple paragraph defining the requirement, teams of programmers need to have detailed sets of requirements that define project goals.

The degree of definition of a set of goals for a large project depends on how well the people who will meet the goals understand the problem. If a mathematician needs a program to solve a problem, there is no gap in understanding if the mathematician knows how to write programs. Difficulties arise when those who propose the requirements don't understand programming and the programmers who will do the job don't understand the goals. This situation is prevalent in the design of control systems, but also happens with complex business or political systems. Multiple millions of dollars and years of time can be wasted when the project has exceeded time and budget limits without producing anything useful. This situation requires project managers who understand both the goals and the programming efforts. Organizations and methods have evolved to deal with these issues, but problems persist.

Programmers

The kinds of programs have become so specialized that one person cannot perform all of them well. The following is an overview of the fields, the associated job titles, and examples of programming languages required.

A program is called an application when it performs some useful task, such as controlling a process temperature or calculating a spreadsheet, or serving as a word processor. Job titles for application programmers include computer programmer, software developer, coder, and programmer. The programming languages that might be used include C, C++, C#, Java, and even Visual Basic.

If the word "analyst" is included in the title, the programmer must be able to understand the user's needs and balance them with the computer's capabilities.

Artificial intelligence may be used in a number of ways, from determining stress in a voice to making decisions based on the knowledge of experts. This field attracts computer scientists as well as software developers who can use languages such as LISP, AIML, Prolog, and the various versions of C.

Database development and maintenance involves different skills than application programming, such as the ability to organize data about real things into data fields and keys. Object-oriented programming was developed to represent the real world as objects of data. People who work in this field are called database administrators or database engineers. Languages include DBASE, SQL, FoxPro, and others designed for a database vendor like Microsoft, Oracle, or Objectivity/DB.

The field of program development for interfaces to hardware requires the ability to understand the workings of the hardware as well as the interface to other software. The job title might be embedded systems developer. Many programming languages have no ability to talk to hardware, so the field narrows down to assembly and C.

Web page developers have to be able to think graphically to handle the graphic web page interface to users. Here we find web designer, webmaster, mobile applications developer, and graphic designer. Languages include CGI, HTML, Java, JavaScript, Perl, PHP, and XML. HDML is used for hand-held devices.

Additional specialties include systems, networks, quality assurance, and security. These fields may involve more management than programming.

Programming

In the early days of computing, when machine names ended in AC for automatic computer, the computing behavior was determined by removable patch panels that wired computing elements together. The panels were removable so that a program could be set up and left in place when the panels were changed to get different behavior. Toggle and rotary switches were also used to set up circuits. This led John von Neumann and others to design stored program machines that used bits set in various kinds of memory to determine the machine's behavior.

Setting the bits in memory got tedious, so programs called compilers were developed that could translate text that followed a set of rules into the bits that could be interpreted as instructions or data. The compiler could detect violations of the rules caused by mistyping or impossible operations, but could not detect logic errors that would cause the program to fail when it was executed. The wonderful thing about programs is that the errors can be discovered and corrected until the program gives the correct behavior for the assigned task, such as sorting a list of names alphabetically or calculating the trajectory of a projectile.

The first compilers were called assemblers. They used an assembly language created to give names to instructions and memory locations, and to allow control of the flow of the program based on conditions tested when the program was running. When single programs became too large, assembly language was used to write routines and subroutines that performed common computing functions. Since the routines were parts of a larger program, compiling the routine produced a binary file called an object. The programmer then wrote a script to tell a program that built the main program from its routines how to link the object files together to create a binary machine language program that could be executed.

Higher level languages were created to avoid dealing with machine instruction languages such as assembly language. For example, an assembly language program must explicitly load the accumulator register with the contents of a memory location, load another register with another number, explicitly add the register to the accumulator, and then store the result in a memory location. This requires about four lines of assembly language code, as follows:

```
lod acc, X   // Load the accumulator register with
                the contents of X

lod R2, Y    // Load Register 2 with the contents of
                address Y

add R2       // Add the contents of R2 to the
                accumulator

sto acc, X   // Store the results back into address X
```

Using a higher level language, one writes:

```
X = X + Y;   /* Add Y to X */
```

In this example // and /* tell the compiler to ignore the comment that follows.

The C Programming Language

What follows is not intended to teach you how to write programs in C, but to give an idea of what goes into writing programs. C is a very popular and powerful language that is relatively easy to learn, but it does not have a graphical interface.

From the preface to "The C Programming Language" by Kernigan and Ritchie, 1978:

> C is a general-purpose programming language which features economy of expression, modern control flow and data structures, and a rich set of operators. C is not a "very high level" language, nor a "big" one, and is not specialized to any particular area of application.

Your first hint that you are dealing with a book on programming is that the Introduction is Chapter 0. Programmers frequently deal with lists or arrays of data by giving a name to the first item in the list and referring to other items as name plus an index. An index of zero gets the first item. The tenth item is name + 9. This unnatural convention leads to new programmers being "off by one" when they check the results of their program.

Chapter references in Kernigan and Ritchie will be preceded by KR to prevent confusion.

KR Chapter 1, "A Tutorial Introduction" gets the reader started by explaining how to write a program that will print "hello, world" on the default printer. This requires learning how to create a file that can be compiled, how to compile it, and how to execute (run) it. There is some magic in seeing a printer do what you have told it to do by typing a few words in a text file. The magic only grows as the programs you create get larger and work the way you want them to. There will be times when version one doesn't work, and you will expand your knowledge by finding out why it doesn't work. After much practice, you will be ready to write more complex programs.

KR Chapter 1 also introduces program flow control with "for" and "while" loops. A loop is a section of program code that repeats until some condition is satisfied. For example, reading from some input or storage device into an array in memory until a carriage return or other end of file mark terminates the input.

KR Chapter 2 covers data types, operators, and expressions. Data types tell the compiler how many bytes to allow for a variable when it is used in the code, such as one byte for a character, two for an integer, and four for a floating point number. There are ways to tell the compiler to modify the rules, using "long," "short," and "double." The arithmetic operators are the familiar +, −,

*, /, as well as % (modulo). The relational and logic operators include greater than, less than, and, or, and not.

Expressions are combinations of variables and operators. If the expression assigns a value to a variable, it has an equals sign and may be called an equation. There are rules for the order of precedence of operators in an expression that can cause great trouble if they are not understood. For example, * and / are evaluated before + and −, so that 1 + 7 * 3 or 7 * 3 + 1 both equal 22 since 7 * 3 is evaluated first. Parenthesis have the highest precedence, so that 7 * (3 + 1) will equal 28. All of these rules, many more than are shown here, must be memorized in order to raise the probability of having your program run correctly the first time. There is a reason why this material is presented early in Kernigan and Ritchie.

KR Chapter 3 presents ways to control the flow of your program, now that you know how to manipulate data using expressions. First, the C compiler ignores white space in lines of code, including the spaces and tabs that make your code readable. The semicolon terminates a "statement" that normally contains at least one expression. Braces (curly brackets) begin and end a block of code, such as a "while" loop. Using blocks of code can produce a "structured" program that avoids the use of "goto" statements that can make it impossible to follow the program flow when reading a program. Flow decisions can be made with "if," "else if," and "else" statements. These are statements that begin with one of those choices followed by an expression that evaluates to a logical true or false value. Braces may be used to block code under each choice. The "switch" statement eliminates numerous "if" tests when a single variable is being tested to make a unique selection. Following "switch (variable)" is a block of statements that each begin with "case" followed by a constant that could be one of the values of the switch variable. When a match is found, statements following that case are evaluated. These statements must end with "break" to leave the switch code block and continue with the program. Loops are also covered in more detail than in Chapter 1.

KR Chapter 4 introduces functions and talks about ways to structure programs. Some people write one long program to accomplish a complex task. This leads to code that is difficult to debug, even with comments. A functional specification for a complex task usually reveals individual bits of functionality that can be defined. It is better to break the program up into many functions, which are small groups of code each with a clearly defined purpose that is easily tested. Often, a function will be used in several places in the main program. Some languages would call them subroutines, but C doesn't have subroutines, just functions. In fact, C compilers come with a library of common functions that a programmer may use, but new functions can be defined for a program, taking care not to use library function names.

KR Chapter 5 discusses the use of pointers. A pointer is a variable that contains the address of the memory location containing another variable. The use of pointers gets interesting when you use them to do address arithmetic.

The following block of code illustrates simple pointers and the use of the & and * unary operators.

```
int x, y;      /* Create integer variables x and y */

int *px;       /* Create px as a pointer to an integer
                  variable */

px = &x;       /* The address of x is assigned to px,
                  so px points to x */

y = *px;       /* The value in the address pointed to
                  by px is assigned to y */

y = *px + 1;   /* y is assigned a value of x + 1 */
```

An integer has a certain size in memory, perhaps 16 bits, but a pointer has the size of a memory address, which may be larger than an integer. The compiler has to know the size of the variable pointed to by a pointer because it will read or write the number of bytes in memory that start at the address given by the pointer. A programmer using pointers must be acutely aware of what is happening in memory, which requires a talent that is denied to many people.

KR Chapter 6 is about organizing data into structures. The standard example of a structure (a record in other languages) is a set of payroll data for an employee, such as name, address, phone number, pay rate, remaining vacation days, etc. A better example for this book is the set of data that parameterizes a control function block, such as tag, description, input, setpoint, output, mode, and tuning constants. The declaration would look like this:

```
struct pidbasic {

   char tag[32];

   char descriptor[32];

   float input;

   float setpoint;

   float output;

   int  mode;

}
```

This defines a generic structure named pidbasic because it is just the set of basic values associated with a PID control function block. The declarations within braces define the members of the structure. The names of members do not conflict with any other variable names in a program. The tag and descriptor are arrays of up to 32 characters. The key variables associated with a PID block are all floating point numbers. The mode (auto, manual) is an integer here for simplicity.

That wasn't so difficult, and the usefulness of structures is clear when the same structure is used many times. KR Chapter 6 goes on to describe arrays of structures and recursive structures, which go deep into the weeds. Recursion is the use of a function within a function, resembling what you see when you look into a mirror in front of you at a mirror in back of you.

When you finish KR Chapter 6, you are only 143 pages into a 228 page book, which is a model of brevity as far as compiler manuals go. There are no standard architectures for computers, due to the need to compete, so the way C compiles in one machine may be quite different in another. This is especially true of the microprocessors that are embedded in control systems. Then there are all of the different languages, so you can see why it is necessary to specialize.

Some Simple Programming Examples   Beginning programmers often are asked to write a program that can determine if a number is prime. Here is a simple solution:

```
/* Determines if the number n is prime.  Returns 1
   for yes, and 0 for no. */

int isprime(int n)

{

   int d;                  /* Divisor to test */
   if (n < 2)              /* Is n smaller than the
                              smallest prime? */

      return 0;            /* Yes, n is not prime. */
   if (n == 2)             /* Is n the only even prime? */
      return 1;            /* Yes, n is prime */
   if (n % 2 == 0)         /* Is n evenly divisible
                              by 2? */

      return 0;            /* Yes, n is not prime */
   d = 3;                  /* Divide by odd numbers >= 3 */
   while (d * d <= n)      /* Loop while d squared <= n */
   {

      if (n % d == 0)  /* Is n evenly divisible by
                              d? */
```

```
        return 0;   /* Yes, n is not prime */
    d = d + 2;       /* Try the next odd divisor */
  }
  return 1;          /* No divisor exists, so n is
                        prime */

}
```

The classic program to demonstrate recursion is N factorial. A factorial of integer N is N times each of the integers less than N down to 2. It defines the number of permutations of N objects. Note that the last line of the program does not actually return until `nfact` has been called n times.

```
/* Find n factorial */
int nfact(int n) {
    if (n <= 1)       /* 0! and 1! have the value 1 */
        return 1;
    return n * nfact(n - 1);   /* Otherwise N! = N *
                                  (N-1)! */

}
```

Visual Languages

There were no graphic user interfaces in the early days of microprocessors. As processing power increased, it became possible to have the user interact with a display of functionality rather than use a command line. Users liked the concept and so it grew.

Developments at the Palo Alto Research Center (PARC) for Xerox led to the concepts of windows, menus, radio buttons, check boxes, icons, and pointing devices around 1980. Pointing devices existed prior to mice as hand-held photomultiplier detectors that produced a pulse when the deflected electron beam spot went by on the display. They were used for air traffic control and locating military targets on radar screens.

Programming a Graphic User Interface (GUI) involves placing everything on the screen at some pixel address in the array of pixels that is defined by the size and resolution of the physical display, which was once a CRT and is now a flat array of liquid crystals or tiny LEDs. Programmers found this to be too much like assembly languages, so "what you see is what you get" (WYSIWYG) programming tools were developed.

The concept of relocatable windows for displays made the addresses relative to the location of the window rather than the physical screen, but the addresses had to be translated to physical pixels. This difficulty further drove WYSIWYG tools. The features available on a GUI were known as "chrome," which explains the name of an operating system developed by a well-known search engine company.

There are now many visual programming languages that allow programmers to specify locations with a pointing device rather than writing lines of code that compute pixel positions.

The touch screen has added a new dimension to the GUI concept by allowing users to specify locations without any other pointing device. Touch screens have been in development for many years, but have only recently made tablet computing possible. At one time, DCS vendors wouldn't use touch screens for fear that a fly landing on the screen could act like an operator entry.

Software Development

It is not enough to write a program if that program will be used by other people over a period of years. There is a life-cycle for a program, as there is for many developed and engineered solutions. It begins with a set of requirements that are understood by the users and the lead software engineers. This is translated into a specification that will be understood by the people that produce the final software. A large project will require an architecture design that serves as an infrastructure for decisions about the smaller projects that will make up the final solution. Design documents and test plans are required for each software module. A debugging process is used to correct flaws found by testing (or by customers). Maintenance is required to fix bugs after the design team has moved on to other things.

For more details, see the Wikipedia articles on Software Development and Software Engineering. It can be useful to read the Talk pages as well as the Article, identified by tabs at the top of the fist page.

Software quality assurance reduces the need for quality control by testing, as it does for any project that begins with a design phase. Many of the bugs found by testing are the result of misunderstandings in the design stages. This can result in a design feature being called a bug by the user.

One method of assuring quality is called the V-Model, which may have originated in 1982 at Hughes Aircraft as part of a proposal for automated air traffic control. One of many variants is shown in Figure 8.9. The process begins with the creation of a user requirements document. The process falls apart if the system is so new that the user is not able to specify the requirements, but that is not true for most automation systems because they are based on something that is already being done by people.

Meetings are then held with users and software developers to turn the user requirements into a functional specification for the project. The dotted line in Figure 8.9 separates user knowledge from developer knowledge. The functional specification can be understood by both groups with some cross-education.

The software developers then produce the specifications that they will use to create the software. The specifications are divided into modules if the project is large; otherwise one coding specification is enough. A user would generally need a lot of education to be able to understand a software specification.

With the specifications well understood, it is possible to write code that

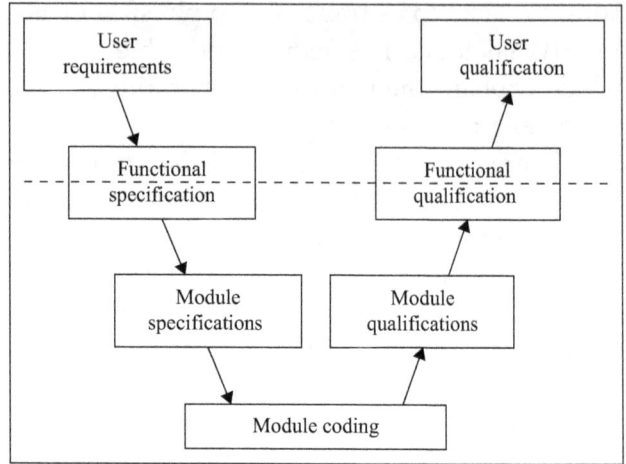

Figure 8.9. V-model for software development.

will not need a lot of debugging by the users. The created code is put into a source code control application that allows many developers to work on parts of the code without interfering with other's work. Some projects use code reviews to assure that the coder's work can be understood by other developers. Part of the work of coding is to eliminate errors reported by the compiler and the module loader. The finished code is able to run on the target system without crashing.

Now we go up the other leg of the V in a series of tests that are designed to show that modules meet module design criteria, the assembled system meets the functional specification criteria, and finally the user tests the system against the original requirements. Each phase of the V model is designed to minimize the consequences of a software defect before it becomes lost in the full system.

Once the system is accepted by the user, then change and configuration management become necessary to keep track of changes made to the software. Changes occur as bugs reveal themselves or users succeed in getting enhancement requests approved.

Programmers are not bad people; it's just that solutions created by programmers tend to vary between impenetrable and opaque. They think in the shorthand notation of the programming language being used, which has to make the result cryptic. Engineering has the concept of a black box, whose contents are unknown to the observer. Properties of the box can only be discovered by stimulating inputs to the box and observing the behavior of outputs. Sometimes there is a manual that describes the intended behavior of the box, but something always seems to be left out. Things that get left out include items that are such a common part of a programmer's environment that it is inconceivable that the user wouldn't already know that. This is not all bad: users that have experienced hours of frustration may know the joy of discovering the stimulus that causes the proper response.

References

Kernigan, Brian W. and Ritchie, Dennis M. *The C Programming Language.* Prentice-Hall, 1978.

Love, Tom. *Object Lessons: Lessons Learned in Object-Oriented Development Projects.* SIGS Books, 1993.

Summerfield, Mark. *Programming in Python 3.* 2nd Ed, Addison–Wesley, 2010.

Engineering

Scientists study the world as it is; engineers create the world that has never been.

—Theodore von Kármán (1881–1963)

Engineering is treated with disdain, on the whole. It's considered to be rather boring and irrelevant, yet neither of those is true.

—James Dyson (1947)

Introduction

Engineers are people who apply their knowledge of how things work to building new things. This has been going on for a very long time, but the word "engineer" goes back to the twelfth century when it was applied to people who built military engines of destruction. As the knowledge of building bridges and towns grew, engineers were divided into military and civil classifications. It was not until the nineteenth century that chemical, electrical, and mechanical classifications were added to civil engineering to become the four basic fields of engineering.

The sources of an engineer's knowledge begin with training and grow with experience—both personal and the described experience of others. Training is done by those who teach, at any level. A recent Nova program "Ape Genius" concluded with the idea that the fundamental difference between chimps and humans was that chimp cultures didn't include teaching. Chimps can be taught, and they can learn by watching others, but they make no attempt to teach others what they have learned, nor do they praise others for learning something.

An important aspect of teaching is teaching what not to do. Scientists may provide new facts and mathematical descriptions of physical laws, but only the experience of doing something that failed provides knowledge of things that should not be done. Henry Petroski wrote "To Engineer is Human" (derived from "to err is human") to drive the point home.

Human culture would be impossible without teachers. They make it possible for each new generation to know more than the one before it, increasing our knowledge of how things work. It is obvious that not everybody absorbs knowledge of how things work, but many get smarter in other directions. Plainly, all brains are not alike. If they were, universities wouldn't have such a large number of paths to a degree. The people who seem to know how things work become engineers if they are driven to get through college by the need to create. This chapter is about those people.

Evolution of Engineering

Considering that humans have been evolving for thousands of generations, and that the tools were extremely primitive for most of that time, where did engineering come from? The ability to engineer your environment has survival advantages, but why did it take so long for engineering to become an identifiable human capability?

A person who built a lasting shelter and taught others how to do it could be called an engineer, but there was no need for permanent structures during the hunter–gatherer years. It would be a stretch to call a stone tool maker an engineer. So it took agriculture with its need to farm one plot of ground for years to create the need for permanent structures for shelter and storage. Agriculture is less than ten thousand years old. Surplus food from agriculture caused the population growth that provided larger groups of people to support engineers with raw materials and demand for products.

There are those who say that natural selection can only apply to individuals, but that leaves traits like altruism with no foundation. Clearly, a group gains a survival advantage from the interactions of the people in the group, and this can lead to specialization of brains into such areas as followers, soldiers, leaders, and artisans. Like individuals, groups that do not survive do not pass on the genes of the individuals in the group. Tribal knowledge evaporates when the tribe is wiped out by barbarians. We have evolved some complex social behaviors and specializations. Engineering is only one of them.

If engineering aids survival, why aren't we all engineers? One could ask the same question about leaders or others who create solutions to life's problems. The answer seems to be that we'd get in each other's way so that no decisions would be made while all of the different viewpoints were debated. The fact that all groups have one leader (or a very small number of specialized leaders) while the rest more or less follow suggests that arrangement to be the one that works best. The politics of group behavior can be quite time consuming for someone who is trying to think of new ways of doing things. It's a rare creative engineer who is also good at politics.

There are many examples of ancient engineering that have passed down to us as artifacts, ruins, or rarely as drawings, proving that there were talented individuals in the population. When, in the fullness of time, the Industrial Revolution began requiring machines to harness energy from greater power sources than man and beast, there was no shortage of engineering talent to do the required invention and construction. Books and schools converted latent talent to reality. Scientific knowledge distributed by those means allowed engineers to more realistically estimate the sizes of parts that would build a lasting creation. Fewer bridges and buildings fell down, and fewer boilers exploded at a time when there were more of them every year. Less time and money was expended on ideas that could not work. Nevertheless, people can be found with designs for perpetual motion machines today.

Prerequisites

It all begins with curiosity about the way things work. In this sense, engineers are born, not made. The curiosity has to be innate and then nurtured by experiences to make it grow. The building blocks, chemistry sets, electronics kits, and mechanical assembly kits all feed the curiosity about how to build things with these parts. Kill the curiosity and the brain will form different neural paths away from creativity. You'll know that your child is a born engineer if construction projects emerge that weren't in the instruction book. If the first project in the book wasn't completed, seek another direction than engineering. Remember that children don't have the fine motor skills to build something with small parts (or play the piano) until about their eighth year unless your child is truly remarkable.

Talent is something you are born with and develop as you grow, unless fate prevents it. Each field of knowledge has some shorthand notation that it uses to quickly express ideas. Music has the staff notation, which people who understand it can read and translate into sound on some musical instrument. People with musical talent can look at a sheet of music notation and hear it. This has degrees of talent. A pianist can look at a sheet and hear how the piano would sound playing it. A conductor can look at a sheet of music and hear the orchestra. Talented mathematicians can look at a page of equations and know their truth. Talented engineers can look at schematic diagrams and know how the building will look or how the chemical products will be produced or the purpose of the product or the feasibility of the control scheme. Scott Adams, with his unique insight into the engineer's mind, created "The Knack" to express the difference.

In case you can't find "The Knack" on the Internet, here is a synopsis:

> Young Dilbert's mother takes him to the doctor because he is not like other children. He had taken apart a clock, a television, and a stereo. The doctor expressed the opinion that this was perfectly normal. His mother went on to say that he had used the components to build a ham radio set. The doctor, now concerned, says he would normally

run an EEG test but the machine is broken. While the adults talk, little Dilbert goes over to the EEG machine and fixes it. The doctor, seeing this, turns to his mother and says, "I'm afraid your son has the knack." The knack is a rare condition "characterized by an extreme intuition about all things mechanical and electrical and utter social ineptitude." His mother asks if he can live a normal life. The doctor replies, "No, he'll be an engineer." She sobs.

Creative work in any field requires a memory that can store many facts, and then combine some of them in new ways. Sometimes this is best done by the subconscious mind (whatever that is) and presented in a vivid dream, or a sketch that emerges from a doodle, or sitting and writing a document, or even taking a bath or shower.

The brain finds it difficult to learn facts that don't connect to facts already known. Anyone who has tried to present a new idea to a committee of people who never thought that way before understands the ground work that must be done to get the members up to speed and able to discuss the merits of the idea. A child acquiring knowledge for the first time must be carefully prepared for what comes next. This is the way education used to work before political correctness became more important. One does not jump from the multiplication tables to calculus without a few years of algebra and equations in between. A person that does not have the necessary knowledge structures in the brain will not be able to fit new knowledge into the existing structures. It helps if the ability to build knowledge structures has been developed since a child could understand the necessary words.

Conversely, spending large amounts of time self-absorbed in social media can produce a person who is only fit to respond to advertising. This is why advertisers spend large amounts of money on social media.

Potential engineers like to take things apart to see how they work. This is a great way to learn successful design and manufacturing techniques. Potential successful engineers can reassemble what they took apart and make it better. Engineers are not necessarily neat and tidy, because they may need to refer to many sources to refine a design. A new boss felt the need to require his engineers to have clean desks, possibly because of some magazine article. The engineers complied, but their productivity was reduced to a third of what it was.

Engineers must communicate detailed information to other engineers. This almost always requires drawings, from sketches on napkins (or tablecloths) to the product of trained detail draftsmen. Computers have eliminated the large open rooms of engineers and draftsmen working at large, tilted tables with vellum, pencils, templates, French curves, and electric erasers, but the need for drawings remains. Each engineering discipline has its own drawing formats, from control block diagrams to detailed assembly drawings with an attached bill of materials.

Engineers creating something new try to model what they are doing as realistically as possible, and test the model. The model is constrained by what is known, so new knowledge may need to be developed. Consider NASA, which has continually pushed at and defined the edge of space technology. We did not go to the moon with the first version of the system. Incremental steps were taken, including orbiting the moon without landing to see what surprises occurred before trying the heart-stopping descent that ended in "The Eagle has landed."

Finally, it isn't all about creating something new. An existing system may develop a problem that requires a solution. This requires diagnostic skills that are heavily dependent on past experience modified by the situation at hand. Engineers are driven to find a solution, as the following short story illustrates:

> A lawyer, a doctor, and an engineer were scheduled to be the first to be executed in a reliable manner by the Guillotine in 1792. The lawyer was strapped to the tilting board and tilted so that his neck was under the blade. The executioner pulled the lanyard, but the blade didn't fall. The executioner placed the lawyer on his feet and said, "You are free to go. God has pardoned you." The same thing happened to the doctor. The engineer was positioned, but struggled to look up at the blade. He said, "I think I see the problem."

The Engineering Environment

Creative engineers would be happy to create in an environment that is not limited by the needs of other people. This is one reason why teams are not noted for their new ideas. Teams are created by management to solve problems. At times, the team is no better than its weakest member. If the team has more than six people, communication difficulties will limit its progress.

Aside from the team, engineers have to work with people who really don't understand what they do, such as the people who control their professional advancement and salary increases. It is a fact that the Human Resources department was invented to enforce policies that prevent a manager from saying that all of his people are excellent. They have to fit the curve, whether that is realistic or not.

One of the best books to explain the relationship between engineers and management is "The Dilbert Principle" by Scott Adams (1996). Understand that managers, like other people, follow the rule that no two people are alike. There are good people out there. Adams makes his living by finding the worst cases. The problem is that it is so easy to find dysfunctional management.

Greg McMillan started a series of books specific to control engineers with "How to Become an Instrument Engineer." There are many examples of well-meaning engineers coming up against people who manage to prevent good things from happening in a logical order. Adams dwells on cubicle dwellers while McMillan and others describe plant startups. It is hard to imagine that an engineer could be found that only had flawless startups. McMillan and also Greg Shinskey have written some very good books for fluid process control engineers. Their books will be valuable for as long as PID is a common control algorithm.

Automation Engineers

There are a great many fields of engineering. The Industrial Revolution gave rise to the fields of civil, chemical, electrical, and mechanical engineering. The discoveries of scientists in biology, atomic physics, and astrophysics, to name a few, have created many more. The engineering fields are all concerned with creating new things, from nanomachines to space propulsion, with branches into biomedical fields and robotics. We have long passed the time when Michael Faraday could make his discoveries in electricity known without using mathematics.

Automation is focused on manufacturing. Manufacturing is focused on cost, given that manufacturing facilities are funded by stock markets that demand fiscal growth. There are people that teach engineers that a real engineer is one who can do for one currency unit what any (expletive deleted) fool could do for two. In reality, many engineers seek perfection, polishing their work product until it has no flaw. This has earned them the enmity of top management, who feel that good enough is adequate. Perhaps you have heard the saying, "It's time to shoot the engineers and start making products."

A versatile automation engineer must know a great deal about manufacturing processes and their control. The three pound brain limit prevents knowing it all, so people now specialize beginning with the process type. Batch and continuous engineers must know everything about PID control, which is not easy to do considering that new things are still being discovered. In some fields, knowledge has a half-life of five years (see Arbesman in the references). Discrete process engineers must know what's new in sequential or procedural control. All automation engineers must know what is available in sensors and actuators for their field. Discrete control engineers are also required to know about robot capabilities.

A. E. van Vogt wrote a novel called "The Voyage of the Space Beagle" in 1950. The "Beagle" was derived from Darwin's ship and his voyage of discovery. The Space Beagle contains several groups of people from different fields of science, who are sorely tried by the wonders they find in space. This would have been just another SF novel except that van Vogt introduced the concept of a Nexialist—a person who knew just enough about the various fields to be able to see ways to combine them to solve problems. The solutions, of course, are dramatic. Boring novels don't sell. This novel explained the problems with specialization and offered a solution. Search for "Nexialist wiki" to learn more. The amount of new information over the last sixty years, perhaps a million times more, makes it unlikely that a human Nexialist would actually know enough to be useful.

Engineering at Rosemount Control

The early days of process control with computers provide the background for the story of a short-lived distributed control system. This story is intended to describe the engineering that is required to design, prototype, and manufacture a process control system. Engineering does not happen in an isolated space surrounded by infinite resources with infinite time to produce results. Corporate considerations must also be included. Systems have no value if they can't be sold and serviced.

You should know this about stories that span decades: human memory isn't perfect. Even if it were, the story can only be told from one viewpoint. Corporations have many people with many other viewpoints, which can be quite different from that of an engineer. Here, events before 1990 are described with the aid of research.

The story can't be told without occasionally speaking of the systems in glowing terms, but there is no attempt to sell anything because the systems described here are no longer being made.

Rosemount Engineering began in Minnesota in 1956 with a small group of people who designed sensors for aircraft, including pitot tubes for airspeed and accurate air temperature sensors for air density and icing conditions. As the company grew, new engineers developed new sensors that found value in industry for measuring flow, temperature, and pressure. The industrial devices passed the aerospace sensors as a source of income sometime in the mid 1970's. The company's rate of growth attracted acquisitive people who profited by merging companies. Rosemount agreed to be acquired by Emerson Electric Co. in 1976 because the Emerson style was to let acquisitions remain relatively autonomous. Only top management was affected by the change of ownership. No jobs were lost and growth continued.

Rosemount was led by Vernon Heath, who had succeeded the founder as President in 1968. There are corporate leaders who would be glad to replace their people with robots, but Vern wasn't one of them. His legs had been weakened by polio in his youth, which may have given him an appreciation for the people who helped him to survive that dread disease. Vern cared about his employees, and they knew it. Cynics said he did it to avoid unionization, but the man was a gentleman from the beginning. Not so gentle that he couldn't control those in upper management that wanted his job, though.

Many companies have reserved parking spaces for the officers, but not Rosemount. The location of a parking space really matters during a Minnesota winter. All employees were ordered by seniority such that the employee with the most years of service had a sign on a space closest to the employee entrance, and so on out to the unmarked spaces for people with less than ten years of service. Handicapped employees, who were never turned away because of a handicap, bypassed the seniority system to get spaces by the door. Add Minnesota Nice and it was a wonderful place to work. Engineers seldom left for other jobs.

Rosemount got started in industrial process control during the sixties because their resistance temperature sensors were better than thermocouples. At the time, all industrial temperature controllers were designed for thermocouples. There was a market niche for the new sensors, and Rosemount moved to fill it by acquiring Plug-In Instruments of Nashville, TN. Being able to meet the customer's needs for temperature control boosted sales of sensors and controllers. The major market was plastics extruders for fibers and film. This set a business case precedent for combining control with sensors.

In the mid-sixties, Rosemount began hiring people from process industry vendors in order to grow into the industrial market for pressure and temperature sensors. This resulted in the model 1151 pressure transmitter, which began to replace force-balance transmitters developed for pneumatic control and electrified with precision solenoids. Two of the people hired, Charlie Smoot (see the Control chapter of this book) and Vin Tivy, encouraged the development of a control system to complement the transmitters.

Diogenes

Rosemount had opened a branch office in England at Bognor Regis, 55 miles south of London, in 1960 to meet the needs of Europe for aerospace. Romilly Bowden joined the development group in 1967 and filled the need for someone who could design and build a control system. He had ideas for a simpler system than the computer control systems springing up at that time. It's fair to say that all of those systems were designed by computer scientists who saw no problem with asking their customers to understand computers as well as they did. He wanted a system that hid the computer from the user instead of flaunting it. Romilly is mentioned by name because he became the father of the Rosemount control division.

Romilly built a computer from scratch using chips available in 1968. The arithmetic unit had a one bit accumulator since its instructions were processed in a serial manner. The program was hard-wired into a matrix of diodes. The user programmed the control schemes by connecting pre-defined control functions with wires that ended in single pin plugs, analogous to connecting pneumatic instruments with copper tubing. There was no fill-in-the-blanks computer configuration. The user needed to know how to connect control functions to build a control loop, but control engineers have to know how to do that in any event. The wires were plugged into a board called a Channel Function Matrix, which was mounted in a drawer in the standard relay rack that enclosed the system.

Vin Tivy named the system Diogenes, possibly because it had so many diodes. I like to think it was because Diogenes looked for simpler solutions as well as honest men. The system was released for sale in 1972—the same year that the microprocessor that would make minicomputers obsolete entered the market. Production was done in Minneapolis using a Texas Instruments model 960A minicomputer, heavily used by the telecom industry for its reliability.

Today we are used to doing operating system updates monthly. Diogenes had a hardware read-only memory that could only be changed by physically changing the memory board. Tiny wires were threaded through E shaped magnetic cores to provide the fixed operating system. Once wired, a board containing ferrite strips that closed the cores was fastened into place. There was no way that the computer could lose its operating system.

Another unique feature was the Process Interface Module (PIM). In that time of transition between pneumatic instruments and direct digital control, Operations people insisted on having backup control if (when) the computer failed. Each PIM was a backup controller. It converted analog 4-20 MADC inputs into digital values for the computer and converted computer values to 4-20 MADC outputs for the valve positioners. It also contained an analog PID controller with screwdriver tuning adjustments that could take over if signals from the computer stopped. PIMs were mounted in a rack of ten, and had the necessary input and output meters to let an operator see process conditions.

An operator could operate the process from PIMs, but the goal was to use schemes of algorithms solved with precision by a computer to control the process. The operator interface for the computer was a panel with digital read-outs and pushbuttons. The operator pressed a button to select a PIM and then entered digits into a read-out correctly before committing the number to a setpoint, output, or alarm limit value. Seems primitive, but that's what operators were used to using in those days before widespread use of video displays.

Consider the engineering decisions that went into reducing these concepts to practice. A great many details were discovered and resolved in order to produce a working model. Then those details had to be transferred to the production facility in Minnesota. It is one thing to make a device work in the lab and another to make it work in production. The lab can fix a problem by choosing a component that works from a bin of similar components. Production requires many similar parts that are purchased with a specified tolerance. Selecting parts is too expensive, so the system has to work with whatever part the assembler pulls out of the box. This is why development engineers are different from production engineers.

An early system was sold to General Atomic for combustion control in a pilot plant for processing atomic reactor fuel rods. A glowing report of the success of that project can still be found on the web under the title "Process Control of an HTGR Fuel Reprocessing Cold Pilot Plant" by James S. Rode (1976).

Combustion is the most common dangerous industrial process. All fuels that burn as gasses have upper and lower explosive limits. The fuel to air ratio must be kept under close control to stay away from those limits without using too much air. Explosions are caused by a sudden generation of gas whose pressure exceeds the strength of the vessel (or house) that contains the fuel-air mixture. Gasses explode, but high explosives like dynamite and TNT detonate with a shock wave that most vessels can't contain. There was a boiler startup engineer who was missing the lobe of his left ear. He'd lost it when a boiler explosion had separated the heavy name plate from the front of the boiler and sent it sailing past his head. The boiler itself was not destroyed—it just bulged enough to break the name plate loose.

Steam is the most common source of power in many fluid processes, and may be used to generate electricity as the steam pressure is reduced to levels required by, for example, the processes in a paper mill. As such, it is a utility that must always be available. Only the most reliable control systems are considered for utility steam boiler combustion control systems. Diogenes found a niche in those plants that were upgrading from pneumatic control systems and wanted to reduce costs by controlling combustion efficiency with algorithms that ran best on a computer. It helped that the cost of a Diogenes system was lower, and that it had worked well for General Atomic.

Small companies tend to attract multi-talented individuals who feel smothered by the teams in large corporations. One such man hired into the British branch in 1979 and was given the job of installing a Diogenes system in a paper mill's boiler house in Scotland. The chief engineer was a dour Scotsman who didn't take kindly to change. After all, steam was a critical utility, what he had at hand worked, and why would he welcome a new way of doing things? In fact, combustion efficiency could be improved, but the chief engineer placed a higher value on reliability.

The Rosemount engineer would convert a few control loops over to Diogenes during the day, but the chief engineer insisted that they be converted back to pneumatic overnight while the engineer went off site. Of course, no progress was made. So the Rosemount engineer gathered supplies to camp out in the control room for a week, sleeping behind the floor to ceiling pneumatic control panel. He was able to get most of the loops converted to Diogenes in four days, at which point the chief engineer trusted Diogenes enough to leave it in control.

That kind of story does things for sales that no amount of marketing literature could do, so Diogenes sales became more than a blip in Rosemount's financials.

Improved Diogenes

Several years earlier, a group in Minnesota had proposed the development of the Next Generation System (NGS) to follow Diogenes. They chose to locate in Mankato, MN, near a state college familiar to the group leader. NGS would use redundant computers to eliminate the need for the racks of PIMs to back up the process if the computer failed, producing an attractive cost savings. Functional, believable redundancy proved elusive and the schedules started slipping. The distance caused a breakdown in communications that resulted in cancellation of the project in 1977. Upper management does not like to be left out of the loop when they are paying for the project. Truly, it was a case of "out of sight, out of mind."

By 1978 it was obvious that something had to be done to keep up with the competition in process control. Upper management watched the money roll in from sensor sales, and wondered why they had to have a control division. That's when they found out about the user base effect. There were customers who were buying Rosemount products because there was a control system. Users like to get packaged services, so that there is only one call to make when something goes wrong. Upper management had to reassure a very good customer that of course there would be a new and improved operator interface, even though they had intended to cancel the project after the NGS experience.

More subtle forces were at work. Rosemount had to hire people who could sell a control system, and so that investment was made for Diogenes. People who sell systems have the intelligence to be aware of changing situations. It became obvious to them that TDC-2000 and its glass console (introduced in 1975) would make Diogenes look bad, in spite of its high reliability. If upper management did not decide to improve Diogenes with a better user interface, those people would find other jobs and Rosemount would lose its investment. Further, there would be no new income to offset the cost of maintaining the systems in the user base, and the result would be a net loss.

The decade of 1970 was an interesting time for users and vendors, driven by the incredible rate of change of computing technology. Users were forced to upgrade from easily-understood and maintained pneumatic control to computer-based control in order to stay competitive. Vendors discovered that they no longer owned the technology that they used to create products, but had to keep up with changes in computer technology with a substantial investment in engineering talent.

The new user interface for Diogenes was called VDS-25. VDS stood for Video Display System, but the source of 25 has been lost. It was much closer to competitive DCS offerings while retaining the control philosophy of Diogenes. Sales improved, but engineering knew there was a better way to do control. It wasn't orthogonal to Diogenes because that system had many good ideas that appealed to customers. Also, it is difficult to find examples of user companies that discarded their investment in sales and service training in order to embrace something completely different. While VDS-25 was being put into production, specifications were being written for the Son of Diogenes, although no one called it that. Think of the acronym. In development it was not called Rosemount Control System but RCS.

Microprocessors had improved to the point that a separate minicomputer like the TI 960B wasn't necessary. TDC-2000 had shown that distributed microprocessors for controllers and operator interfaces that were connected by a communication network would be accepted by users. TDC also showed that elaborate backup control systems were not necessary because distribution limited the risk of failure of a single computer.

However, TDC and those who emulated it did not have the ability to solve logic problems that Diogenes had evolved in order to do burner management and combustion control. Exxon required a system that would do analog control for a continuous process. Honeywell worked with Exxon because the market for upgrading control oil refineries verged on gigantic. Those pesky logic problems were left to a PLC—except that there was a need for a common operator interface. Diogenes had an edge in the sub-refinery market, and so did its son.

Development of RCS

A new system begins with a series of meetings with engineering, training, and marketing people who have experience using and selling systems. The Diogenes and VDS-25 experience was paired with market studies of what the competition was doing. This produced a list of goals for features without solving any of the how-to problems. Then engineering led meetings to see what was feasible, and to see if any goals could be exceeded or at least redefined in terms of cutting edge technology. In those days, the cutting edge went by in a few months.

Existing systems encapsulated control functions in boxes called function blocks along with the inputs and outputs required to perform that function, and the constants necessary to parameterize the function. RCS decided to use a cube for the foundation rather than a two-dimensional block. A cube had an analog function and a logic function composed of 16 user-configurable logic steps. Each function had a face of the cube for parameterizing the function, connecting the inputs, and setting alarm trip values. The name cube was changed to compblock because the concept of blocks was too widespread.

Diogenes function blocks (originally circles) had status signals in addition to the function output, such as limited high or bad measurement. These signals could be used in three-input logic gates whose output could have an action associated with it, such as freezing the output of a PID function so it didn't wander off when the loop was open because the pump wasn't running. Because the whole control program ran in one computer, there was no difficulty in connecting anything to anything else. RCS was a distributed system with many computers, so the information had to be carried over a network from one compblock to another. The connections were called links, although that term also had a specific use in networking.

The communication system was ingenious and original, at the time. The network was called the Unit Data Highway or UDH. It had two redundant cables running at 1 MHz. A dedicated microprocessor called the Coordinator Processor (CP) was used to handle the communications for ten controller cards, which could each handle 100 control loops. The CP bundled up link messages bound for other CPs and sent them out over the UDH. An algorithm used fifteen statistics about the quality of communications on each UDH, such as the number of retries and confirmation timeouts, to decide which UDH to use. If both had no errors, alternating highways were used. If one went bad (perhaps severed by a fire axe), the CP would still try it occasionally to see if it recovered. Contemporary schemes used only one highway at a time, made a hard switch on failure, and relied on the user to switch back.

The CP also managed non-volatile memory (NVRAM) for the controller cards that it connected to the UDH. NVRAM contained program images for the CP and controllers as well as complete configuration backup for each controller. This allowed a CP and its controllers to reboot after a power outage without any UDH traffic. NVRAM used magnetic bubble memory until it was possible to use battery-backed-up RAM chips.

Distributed control requires calculated information to flow forward, from sensor to PID to output function, in a deterministic manner. A series of control equations are being solved to process a control loop from beginning to end. This must be done sequentially and in a fixed period of time because the equations are tuned for a particular sample rate. In some cases, it also requires information to flow backwards from output to PID so that the next time the PID is executed, it will know whether the output performed the last calculation or not, and why not. This affected the structure of the data in links.

Diogenes used 16 bit fixed point numbers with scaling. RCS had to use 32 bit floating point numbers in engineering units to keep up with other systems. Another 16 bits were added for the data status flags generated by the calculations in each block. The results of the 16 logic steps added the last 16 bits so that a link value was a 64 bit value structure. This was not negligible in a 1 MHz network with possibly 100 links per CP node. Backwards communication doubled the load. Fortunately, few blocks require links to a function in another CP, and if external information is delivered to many blocks in a CP, such as a shutdown signal, it only has to be sent once to each CP.

The UDH used a priority scheme for three classes of messages: links, alarms, and other. A communication cycle began four times a second with a synchronizing tick message (there were no tocks), followed in node order by all of the link messages, followed by alarm messages, followed by anything from messages required to build a screen display to a complete recording of all of the data in all of the controllers. Only links had to finish before the next tick; the rest were queued for delivery as time permits.

That was just the engineering required to get the control job done reliably, and that was perhaps the easiest part because the control functions were well defined after fifty years. Next came the operator interface and the engineering interface. The essential operating information was presented to operators as group displays derived from pneumatic panel instruments (people change their ways slowly) or graphic displays that showed a schematic diagram of the process and populated it with data from measurement and control devices. The users had to build the graphics, so a method was required to allow that to happen.

Release to Production

The heart of the system hardware was called a Control File, which was the metal frame, backplane, and sockets that held the printed circuit boards. A full control file contained two communications interface boards (one for each UDH), two redundant power regulator boards, two redundant coordinator processors, non-volatile memory, and eight controller boards. The original controllers came in three flavors: analog, discrete, or multiplexor. In a later version, a multi-purpose controller could do any one of those functions, and a pair of controllers could be made redundant by adding a jumper module to the backplane.

The Operator Console was initially designed as a pedestal with a desk-level area containing special operator keys (not far from the VDS-25 design) and a four inch diameter trackball, surmounted by a 21 inch color Conrac monitor. Under the desktop was room for a card cage containing the UDH interface, power regulator, processor, video generator, SCSI board, printer interface with real time clock, and NVRAM. The SCSI peripherals were a 20 MB hard drive and a tape drive for loading and backing up the system configuration. That four inch trackball provided the smoothest cursor positioning you can imagine. Later table-top versions used a two inch trackball in a small box that sat alongside of an operator's keyboard and possibly an optional QWERTY keyboard in the more familiar desktop configuration. The pedestal looked good, but users preferred the table-top layout.

The control file and operator console power supplies were powered by custom 30 VDC power supplies with seven amp-hour batteries to ride through brief power failures. Redundant power distribution panels allowed two or more power supplies to provide system power to A and B power busses. The user could provide much larger battery backup if required. The supplies also provided power to the field instruments through their two-wire 4-20 MADC connections, which is why they provided 30 volts and not 24 or 28 volts.

When the system design is ready to transition to manufacturing, the company executives become involved in choosing items that will be part of the image of the company, such as the colors and the name. RCS became Rosemount System 3 or RS3. Three was a popular number for systems in those days. The color scheme was blue to match the color of the pressure and temperature sensors being shipped. Marketing has some freedom with names as well, so the UDH became the Peerway. Peer-to-peer communication had a good reputation at the time.

Four areas of engineering are required to produce a control system. Design architects for hardware, software, and system create the design, and the senior production engineer creates the changes necessary to manufacture more than one system. Manufacturing includes designing the test equipment that detects bad parts and assemblies before they get into the system, as well as the jigs and fixtures required for assembly. The hardware architect has to transfer knowledge of the design to drawings that show production what has to be made, cooperating with the production engineer to define assemblies that can be manufactured.

Production also has to make the disks or tapes that contain the system software. The software architect is responsible for developing software that works as well as controlling the versions of that software when the inevitable defects are found and fixed. If the software passes testing on the system, it is released to production. The more complex the system, the less likely that testing will be adequate to find all of the defects. A defect is a difference from behavior specified in approved design documents. If something wasn't included in the specification, it can only be detected by the customer as a difference from expected behavior. Then there is a discussion of whether the defect is a bug or a feature. Sometimes defects will not show up on a small system, but appear as the system is taken to its size limits.

When the design is released to production, the system architect is the only one who knows how the system is supposed to work. This knowledge has to be transferred to many groups. The marketing group has to create glossy brochures with features and benefits. The technical writers have to write the manuals that customers and systems integrators will use to customize the system for their use. The training department has to prepare course materials for customers and sales training. Field service technicians have to be able to fix problems to keep customers happy. Systems engineers at the company have to know how to configure systems for customers that don't have a systems integrator or an internal engineering department.

The system architect has to write a book that explains how the system works in order to get all of those groups started. Much of the book can be written as the system is designed, and rewritten as design changes are made. Some of the groups participate in the design process and so have advance knowledge of how it all works—enough so that they don't have to start from zero knowledge. But there is a window when the sales process requires someone who understands the system well enough to describe it to a potential customer before the glossy brochures are printed.

Concurrent with all of this, the system architect used the system to discover defects. The list soon passed 400, so it had to be prioritized. This was done in meetings with software engineering and marketing. The defects that all agreed would cause severe trouble for the customer went to the top of the list, followed by less severe bugs, followed by annoyances and finally those that might be features. Of course, there were not enough resources to fix that many problems before release. Version 1 did not get out of the factory. Version 2 went to the first customer, who added more bugs to the list. Version 7 was quite good, but it seemed that the software would never be bug free. Enhancements to keep up with technology also caused version changes, each with a list of bugs to fix. The last version shipped was 15.

Figure 9.1 shows an engineer testing a large RS3 system at the factory in about 1988. One operating position and five control files are shown out of the 15 control files and four operating positions in the system. The capacity of the Peerway was 32 nodes. This was the first time that a system of 27 nodes had been assembled, so test conditions were set up to load the Peerway with an unusual number of messages. The trackball at each operating station used the Peerway so that it could be switched to position the cursor on adjacent monitors, which had individual connections to the Peerway. It was possible to send a burst of so many hundreds of alarm messages that the delay

Figure 9.1. An RS3 system.

from moving the trackball to moving the cursor was five seconds, but nothing crashed and no messages were lost.

There was a remarkable incident about five years after introduction that emphasized the difference between development engineering and production engineering. Customers began reporting that the controller cards were crashing and restarting. There were no process accidents, but that behavior made the operators nervous. The crashes were random, making the fault the worst kind to find. Software was assumed to be at fault, so a test system was set up in the lab and loaded with software that had fault tracing tools enabled. The tools failed to find a software fault. After several weeks of this, it was time to consider the impossible. Hardware probes were put on the microprocessor to see if a logic analyzer would reveal anything. Finally it caught the reset line to the micro in the active state.

Microprocessors that run continuously, as they do in control systems, have a watchdog timer that resets the processor if it stops. The timer is reset by software, so if the software stops executing, the timer will time out and reset the microprocessor. The design used a very common type 555 timer chip. The timer chip vendor changed to a new method for fabricating the chip but kept the same part number. Purchasing for production ordered the same part, but the new ones had a defect that caused it to trigger at long random intervals. Changing to a different vendor solved the problem, but the existing controller cards had to be recalled and fixed.

Initial Field Experience

The first RS3 system was sold to American Maize in Hammond, Indiana. This large corn milling plant had a multiple phase approach to entering the computer control era. The first phase included a starch processing plant that had 22 storage tanks and pumps, and a piping transport header with 167 valves (IIRC). The upgrade included block valves to replace a hose panel, where the operator could connect one tank to another by reconnecting stainless steel hoses on the panel. As anyone who has done this before knows, it is one thing to design the system on paper, and quite another to implement it and make it work. To begin with, the basis for the paper design seldom includes all of the details.

The team sent to the site included representatives from Marketing, Field Service, Training, and the system architect because Systems Engineering was not yet up to speed on system configuration. The man from training proved invaluable because the union representative was dead set against automating the system. He stalked out of the training class, but returned after a few days when his people said there was nothing to worry about. That could have gone another way entirely.

The capability of RS3 to do bit vectors allowed groups of 16 valves to be commanded and tested with one logic step. The ability to do a state machine in logic allowed a starch transfer to have several states. The "Check closed" state made sure that all of the connections to alternative routes were closed. "Open Route" commanded the necessary valves to open, in groups of 16. When this was confirmed, "Start Transfer" started the source pump. When the desired amount was transferred, "Stop Transfer" stopped the pump, and "Wash Route" pumped relatively clean water through the route. "Drain" let the rinse water drain from the pipes, and then the route valves were closed to let another route be established. It took three weeks to configure the system to do 24 different transfers, which was considered fast.

There were some glitches, as is to be expected at a startup. An operator complained of unwanted digital commands, such as opening valves and starting pumps. This was traced to the Field Service person taking pictures of the control card racks with a flash camera. The digital I/O cards each had a microprocessor that had an ultraviolet light erasable ROM on the chip. The flash temporarily confused the ROM, so black tape was put over all of the chips. On the positive side, RS3 power supplies had been designed with battery backup. During a demonstration for another customer, American Maize lost power for several minutes. The only light in the control room was from the operator's console, as RS3 continued to control the process. The operator was able to manipulate valves to prepare for restored power and restarting the pumps.

The first European system was sold to Bowater-Scott in Gravesend, UK. The town's name may come from the last graves dug along the Thames during the Bubonic Plague. The process was a tissue mill with a 14 foot diameter Yankee dryer. The same startup team lived in Gravesend during the startup, joined by people from the mill and from Rosemount England who very much wanted to see the new system.

The Peerway used non-standard triaxial cable connectors. Only four were needed, so only four were shipped. The cables had to be cut to fit the installation and the connectors attached. The smallest part of the assembly was the center pin, so naturally the first one was dropped and lost. They had to make do with one cable until more connectors could be shipped from the USA.

A labor crisis arose when it was discovered that an operator on the third floor had to report the status of a rotating filter as being in one of three states. The system had not been sized with any spare I/O points in order to have the lowest bid. It would be necessary to add another operator's console on the third floor, and training for another operator, which would delay the startup. The system architect found a spare analog input, so a three position switch and three resistors could deliver currents within the range of 4-20 milliamps. RS3 logic steps could differentiate the three levels of current and report them as the three states of the filter. This averted the labor crisis.

Bowater had saved money on reversing relays for motors by using standard relays and letting the control system prevent them from both being on at the same time. When the system was tested on site, fuses blew and contacts welded because two relays were on at the same time, shorting the line. This was traced to a defect in the solid-state pilot relays used by Rosemount. The pilot relay manufacturer was able to correct the defect in time for the planned startup.

Startup of the steam flow control loop for the Yankee dryer revealed a major problem in the system. Override control was required to limit the steam flow while the dryer warmed up. The shell is quite thin and easily cracked by thermal stress. A crack would be a major disaster for both Bowater and Rosemount. The problem caused the steam valve to ratchet open a bit each time noise in the steam flow measurement changed the selection between the steam flow and pressure controllers. It took real-world experience to reveal the problem. Fortunately, there was room to configure additional logic steps to work around the problem.

Finally it was time to make the first tissue. Everything came up to speed smoothly and was running well when a plastic hose fitting in the Fourdrinier section failed and broke the web in a spectacular spray of pulp. Startups are usually interesting times.

ABC Batch

By 1986, most of the knowledge had been transferred and the defect list (called Rough Edges) was under control. Coincidentally, a marketing engineer had transferred in from Europe, bringing knowledge of the work that a German group (NAMUR) was doing to organize batch process control. The decision was made to add batch control to RS3.

Most batch control systems of that era used an external computer that communicated through a supervisory interface to manipulate control loops. RS3 would have batch control processing built into the Coordinator Processor in each control file, to take advantage of the reliability and redundancy already proven in use. This required a processor upgrade and more memory, which were easily accomplished. This capability alone was the deciding factor in several buying decisions by users. No one else had distributed batch control with NVRAM backup.

The software was not so easy. RS3 had a capable programming language that was used to configure the logic steps, which needed some additional features to allow batch control scripts to be written. A framework was designed that allowed a script for a batch operation to use alias names that would be resolved when the script was started. Usually aliases are just used for control loop tags, but these were organized by the control points in a batch unit and the control points in a material source. A script could also use formula values provided by another script if the same procedure could make products that only differed in formula values.

Several other things were required, such as the ability to interact with an operator for manual operations, but the intent of this section is to illustrate kinds of engineering problems, not to list them all.

As it turned out, Foxboro, who had been a major provider of batch control systems, released a new system without making it backwards compatible with existing I/O installations. Rewiring the I/O is expensive, so Exxon went on a tour of batch system providers, intending to end with Provox. They ended up buying an RS3 system, and soon after another was sold to Arco. The growth of RS3 began to impress those in upper management who did not understand why Rosemount was building control systems when there was more profit in transmitters.

The End of RS3

The control division prospered from 1985 to 1999, with many more interesting startups. The thousandth control file was sold in 1998, when the user base was over $100,000,000. Diogenes had only reached five million. There had been signs of trouble ahead. Funds for a major revision of the system to catch up with technology dried up in 1988. It turned out that Emerson had begun due diligence for a possible merger with Fisher Controls. The future of RS3 had become uncertain because Fisher had a $200,000,000 user base of Provox control systems that also needed to catch up. Emerson wanted Fisher's valve business to complement Rosemount's transmitters. There would be one control system for the merged companies. Rosemount's system architect was diverted to ISA standards committees SP-50 and SP-88, followed by Foundation Fieldbus. Rosemount development engineers began drifting away to other jobs. Management could not explain what was going on because preparations for a possible acquisition have to be top secret lest the stock prices be disturbed.

The merger was announced in 1992. The Users Group meeting in September (in Banff) was dominated by the idea that the merged company would not support two control systems and wanted to know which way that was going to go. Exxon had bought RS3 batch control systems, so their concern was palpable. Meetings between the two engineering groups were organized and the corporate jet employed to ferry people between Austin, TX and Minneapolis. The engineers experienced yet more interesting times.

An irreconcilable difference was discovered in that RS3 used on-line configuration while Provox did all configuration to a database, which was then checked and finally downloaded to the controllers. The advantage of on-line configuration is that any change had an immediate effect on the user's system, which could be undone if necessary before it propagated further. It's a great teaching tool. The disadvantage is that you need to have a system to configure. Most systems are configured while the system is being built, so RS3 Systems Engineering developed a way to fill in tables with configuration items that could be loaded into the system when it was assembled for factory test.

Provox did not have a factory Systems Engineering group. Fisher customers were their sales representative (rep) companies, such as Puffer-Sweiven in Texas and C. B. Ives in Pennsylvania (now just Ives and selling Siemens control). The reps wanted database configuration so that they could sell the service of configuring the system before installation. The database had the advantage that algorithms could be used to be sure that there were no obvious conflicts in the configuration. The disadvantage is similar to compiling computer programs. You write the whole program and then submit it to a compiler that eventually returns a list of errors. Sometimes a long list of errors has a common cause, but it takes time to locate that cause. The delay between configuration entry and resulting error can be long enough for you to forget why you made that configuration entry.

That was not the only major difference, as is to be expected from two independent development efforts. The RS3 group knew they were in trouble when they heard that the largest Fisher rep had threatened to buy Emerson if they didn't get database configuration. Further, a seemingly unrelated Emerson executive decreed that RS3 would cease using Unix and start using Microsoft products. Oh, and the focus would be on object-oriented design.

RS3 had been using an efficient eight bytes of data to represent variables. Object-oriented design multiplied that by more than a factor of ten. At a later presentation to Exxon, the Provox software architect described how they would be moving to an object-oriented database by Objectivity, which was quite new at the time. The Exxon representative turned to the Fisher-Rosemount president and said, "John, you have got to take away their magazines." He considered object-oriented design to be an unnecessary complication, when it was the hot new thing in the technical magazines.

While this was going on, the SP-50 work on standard digital field signals bogged down. Fisher-Rosemount joined other companies, including Siemens, to form the Interoperable Systems Project (ISP) in order to start manufacturing field devices to a common communication standard. There would be no endless debates over the virtues of competing communication systems, and it would include basic control in field devices.

Some engineering meetings were held to establish the points of interoperability. Then the architects were sent to a small hotel in the Black Forest in Germany in November of 1993 and told they could come home when the specifications were written. This was also an interesting time. It took three weeks, in an environment very different from an ISA consensus standard development.

ISP became the Fieldbus Foundation (FF) and the rest is history, some of which is in the Communications chapter of this book. There was an unexpected effect on the design of the new Fisher-Rosemount system, which had become quite close to being Son of Provox. Instead of improving the old control algorithms, the system would be designed to operate with control in the field. The system would use the same basic control algorithms, down to the parameter names and behavior, as those specified by the FF. This caused some consternation, but it was simply a matter of programming. Resources were not provided to convert RS3 to FF behavior. The writing was on the wall for those who could read it.

There is no engineering content in the story of the end of RS3, except that the reason given for closing was that it would free up money needed for development of the new system. RS3 was the only system making a profit at the time, but that didn't matter. Three months notice was given on April 28, 1999. A development engineer posted a cartoon outside of his cubicle of a fat dragon picking his teeth with a broken knight's lance and surrounded by bits of armor. The caption was, "Sometimes the Dragon Wins." You can find variations of it on the web.

Follow Up

The RS3 system installed at Channelview, Texas, in 1988 was turned off on March 26, 2013, to be replaced by a new system. Ideas were sought for a celebratory "T" shirt. One of the candidates was, "What Happened to Deltas III and IV?"

References

Rosemount 1956–2006 Fifty Years, 2006, Emerson Process Management [has a picture of Diogenes but nothing about the control division].

Adams, Scott. *The Dilbert Principle.* Harper Business, 1996.

Arbesman, Samuel. *The Half-life of Facts: Why Everything We Know has an Expiration Date.* Current, Penguin Group (USA), 2012.

McMillan, Gregory and Weiner, Stanley. *How to Become an Instrument Engineer, Part 1.523.* ISA, 1994.

Petroski, Henry. *The Essential Engineer.* Alfred A. Knopf, 2010.

Petroski, Henry. *To Engineer is Human,* 1982–1985. Various reprints.

Rode, James S. Process Control of an HTGR Fuel Reprocessing Cold Pilot Plant. General Atomic Project 3225, Oct. 1976. [available in PDF at http://www.osti.gov/bridge/product.biblio.jsp?osti_id=7257816]

Van Vogt, A. E. *The Voyage of the Space Beagle,* 1950. Various reprints.

The Future

We must welcome the future, remembering that soon it will be the
past; and we must respect the past, remembering that it was once all
that was humanly possible.
—George Santayana (1863–1952)

Trying to predict the future is like trying to drive down a country
road at night with no lights while looking out the back window.
—Peter Drucker (1909–2005)

Introduction

We can prepare for the future and possibly shape it if we can foresee
possibilities and estimate realistic probabilities for them. Simple predictions
by extrapolating trends are not useful. Such predictions include the paperless
office, the leisure society, and the flying car. A careful examination of the
details required to support those predictions would have greatly lowered their
probability. When Einstein's equivalence of mass and energy became known
to the popular science writers, they cheerfully predicted ocean liners crossing
the Atlantic using only one lump of coal. The details of the conversion process
were not considered.

Planning for the future should begin with a search for the drivers of
change that we can see today. Drivers will appear tomorrow that we couldn't
see, but that is the best we can do today. Sometimes tomorrow's drivers are
hidden in new science that is only discussed in esoteric journals. It can take
twenty years for new science to become commonplace, but a lot of new science
doesn't make it. If science doesn't become technology, it won't build things for
the future. Sometimes a new way of measuring something makes new science
possible. The microscope had a huge effect on biology.

There are technologies that are waiting for scientific discoveries to make
them practical, such as storage batteries that would make solar power viable

on a large scale. New science and delayed technologies are possible drivers of change. Today, additive manufacturing has the potential to create objects from molecules to engine blocks, or biological cells to organs to meals. There is no shame attached to reading speculative fiction to find future technologies, although the probabilities for matter transporters and warp drives remain infinitesimal. Books such as "Future Science" (2011) may be more useful when they can be found.

When you are planning for the future, be aware that it is only one of many possible futures. In one of them, the value of a nation's resources becomes the wealth of a few individuals who are above nations. That future needs reshaping, because it has always ended badly throughout history.

The term "disruptive technology" was coined by Harvard Business School professor Clayton M. Christensen in his 1997 book, "The Innovator's Dilemma." He differentiated between sustaining technologies that represent incremental changes to existing technology, and disruptive technologies that start off as appealing to a very few early adopters and grow to threaten the companies that were secure with sustaining technologies. An example is the sustaining technology of the personal computer and the disruptive technology of the smart phone and tablet. Mobile device technology was known but impractical until advances were made in batteries, chips, and displays.

It is possible to be biased to the point of blindness by previous experience with a rapidly developing field. An experience with 3D printing just four years ago did not prepare me for what you will read in this chapter. A similar story applies to model predictive control. My experience was that you couldn't keep it running. Evolution has us remember past experiences as if our survival depends on it. Today, technical experiences become out of date if they are not continuously tested.

This chapter will address the future of automation by first discussing the future of manufacturing and then the future of computing. The design of automation systems derives directly from the requirements of manufacturing. The technology used to build automation systems relies heavily on digital computers and their peripherals.

The Future of Manufacturing

It is safe to say that manufacturing will continue to create new products as long as there is a market for them. Markets exist in nations, so the future of manufacturing is tied to the future of prosperous nations. The future of nations will be affected by the growing power of multi-national corporations—organizations that do not care about a nation's balance of trade and feel constrained by its regulations.

A nation cares very much about its balance of trade because that affects its ability to survive. A corporation is primarily concerned with fiscal growth, if it has stock that trades in an open market. A nation has many concerns, but it ceases to be a nation without jobs for its people. The prediction of a leisure society ignored the human need for meaningful work. The future of nations and corporations is outside the scope of this book, but they will affect automation. An interesting book on the subject of globalization is "World on Fire" by Amy Chua (2004).

Nations provide the infrastructure for manufacturing businesses. A nation requires a government to provide a basis for negotiating with other nations, to maintain a monetary system, and to provide for the social welfare of its people. A nation of disgruntled workers is of no use to manufacturing.

Manufacturing needs to ship and receive materials, and obtain utilities such as electricity and water. Nations manage transportations systems and utilities either directly or by encouraging private enterprise. The Internet may be viewed as a utility, along with other means for advertisers to connect with people.

Nations that have national laboratories and organizations like NASA manage technology transfer programs to help businesses to innovate. There may be national services for weather forecasting and precision time. Indeed, there are many more government services that are not listed here, such as defense, banking, and the encouragement of small businesses.

There are at least three major issues with infrastructure that directly affect the automation of manufacturing:

1. The problems of network security must be solved. This may require new operating systems and network interface applications that are not designed for doing market research on Internet customers. See the Security heading later in this chapter.

2. The disparate interface and communications standards that were created by competition must be integrated into a few standards that will serve all users. The "all users" part is difficult, but it should be possible to reduce the number of standards.

3. Training must keep up with the pace of change. Manufacturers must be able to find people with enough training to be able to learn quickly on the job. National accreditation of training is required.

Processes

Manufacturing processes are constrained by changing technology. Discrete assembly machines were constrained to line shafts and cams until a combination of faster PLC processors and advances in stepping motor technology made it possible for motion control to replace the mechanical marvels. The largest advance in fluid processing has been the development of in-line analyzers beyond the sampling gas chromatograph. Batch processes now have the flexibility of computers to replace mechanical drum programmers.

Additive manufacturing   Additive manufacturing, also known as 3D printing, is a recent development that shows no sign of being limited, except by human ingenuity. It is one thing to melt plastic strands to deposit material with an X-Y plotter on a table that drops a small amount after each X-Y pass. It is quite another to spread a thin layer of powered metal and use a laser or even an electron beam to precisely weld the metal particles to the ones that were welded on the last pass.

The method has been scaled up to make parts for an automobile and scaled down to make tools for shaping nanomaterials. Automobile engine blocks have been made, eliminating casting and much machining. Coolant passes in the block no longer have to be drilled and plugged. NASA has "printed' and tested a rocket nozzle for hydrogen-oxygen combustion that appears to be about 5 inches (12 cm) at the exit diameter. Such nozzles have cooling passages that could easily be printed.

The Australians are looking at additive manufacturing, and note that a large printer is not limited to making large items, but can make many small items that will pay for the machine sooner. They also note that 3D printing is cost effective when used to allow frequent design iterations before committing the design to tooling.

Science fiction stories of the fifties had a home matter duplicator as a stock item. It is already possible to use an affordable 3D printer to make objects at home from applications freely distributed on the Internet. 3D scanners are also being developed that will make physical copying possible. Few manufacturers feel threatened by this in its present crude form, but the Wright Flyer was a crude airplane.

3D printers have been used to print organic matter. The emphasis has been on replacement body parts, but the same principles could be applied to steak and potatoes, with a microwave oven section to cook them to perfection. Liquids will need their own dispensers.

Additive manufacturing can be a very attractive alternative to subtractive, such as milling an alloy bar into a part with 90% waste material. Even casting requires machining to achieve precise dimensions. Milling and casting are the traditional methods for transforming the shapes of materials. You could say that welding is the traditional method for additive manufacturing, because the weld fuses the parts together as if they had been cast.

There are things that additive manufacturing can't do, such as create a surface with nothing under it. No diving board to go with a tiny pool, unless you do some assembly to add the board. It is possible to do an inverted pyramid by overlapping each layer just a bit.

Consider that the detail available in an additive fabrication is inversely proportional to the amount of material added at each movement of the nozzle or beam. Casting and milling can create shapes in a fraction of the time required to add tiny bits of material. Additive manufacturing can do some wonderful things, but it can't cause the end of the Subtractive Age of manufacturing. Not even if the speed of adding material increases at anywhere near the rate of the increase in computing speed. There is still the problem of a rough surface caused by printing in layers.

Biotech Manufacturing There is already a significant amount of biotech manufacturing using single cell organisms, primarily in the pharmaceutical industry. Adam Rutherford, in "Creation" (2012) suggests that the future holds promise for more aggressive genetic engineering. While some of the products can be made in bioreactors, some of them require multi-celled animals, such as goats. This puts a whole new meaning on manufacturing. There could be lots of work for production and control engineers, as well as activists and lawyers.

Molecular Manufacturing Continuing down in size, a molecular assembler was proposed by Eric Drexler in 1986. Molecular assembly is done with great ease by RNA in living cells and is used by biotech to make products. Drexler proposed making a molecular assembler leading to mechanosynthesis. The speculators went wild, so Drexler defined nonbiological molecular manufacturing in 1992 as the programmed "synthesis of complex structures by mechanically positioning reactive molecules, not by manipulating individual atoms." Manipulation of atoms is done by nanofabrication, in the dimensions below those of molecules.

Nanomanufacturing There are two aspects to nanomanufacturing. One is the manufacture of nano-sized materials, liquid or solid (or quantum undecided) from bulk matter and the other is the manufacture of parts that have been "machined" to an atomic scale. Nanofabrication refers to the assembly of atoms or molecules.

The concepts have been defined, but the methods are not well known. The US National Science Foundation has formed a National Nanomanufacturing Network for exchange of information that will move techniques out of laboratories and into manufacturing. See http://www.internano.org/

Materials

We are used to dealing with materials that we can see, if only through a microscope. That is changing as new uses are found for nanomaterials, defined as having at least one dimension in the range of 1 to 100 nanometers (nm) (10^{-9} meters). That is about one third of the wavelength of violet light, and so it is not visible. The surface to volume ratio is so high that materials behave in different ways from their bulk forms. Newtonian physics gives way to quantum mechanics. Nanomaterials are not simply small quantities of the materials we know. They have different physical properties: copper becomes transparent and gold becomes soluble in water, which turns red. Inert materials may become potent catalysts for chemical reactions when reduced to nanoparticles.

The upper limit of 100 nm is half the length of the smallest bacteria, so it is below cellular life. It is also below the micromaterials, which still retain most of the bulk properties of the larger materials. The lower limit is four times the size of a hydrogen atom, chosen because nanomaterials are assembled from atoms. We'll need a different name for materials assembled from sub-atomic particles, if we ever get to that level. Creating atoms with particle beam accelerators is not what anyone would call manufacturing—not until they last long enough to be packaged for sale.

Intel is currently making 22 nm parts and planning to go to 14 nm, deep into the quantum realm. The thing about the quantum dimensions is that the probability of some strange thing happening increases. A freshman physics professor (the greatest that ever got chalk on his coat) introduced probability by pointing to a blackboard eraser sitting in its tray. "That eraser has many moving atoms in it. If they all moved the same way, the eraser would jump out of the tray. That does not happen because the probability of all of them moving the same way at the same time is infinitesimally small." There aren't that many atoms in the quantum dimensions. 10 nm erasers would jump out of their trays all the time.

Perhaps the first nanomaterial was discovered in 1985 when a construct of 60 bonded carbon atoms was seen in reaction products. The resemblance to a geodesic dome caused it to be named after Buckminster Fuller as a fullerene. Tubular shapes have also been discovered, made from sheets of carbon atoms one atom thick. The sheets were called graphene. These early materials had properties not normally associated with carbon that had discernable commercial implications, to the point where government labs became involved. It took fifteen years and other discoveries to generate excitement about nanomaterials. Silver nanoparticles were discovered to have excellent antibacterial properties, resulting in new ointments for cuts and even being used in socks to reduce odor.

Things really got interesting when it was discovered that adding a few percent by weight of graphene to some polymers could make them much stronger and able to stand higher temperatures, which caught the attention of the auto industry. Stronger plastic means lighter parts, and a lighter car has better fuel economy. Such parts are available today even though the mechanism is not well understood yet.

This brief summary is intended to show that nanomaterials will be a part of future manufacturing, in ways that are as yet unknown. It isn't all positive, though. Those socks with the silver nanoparticles that killed bacteria lost silver when washed. The silver in the wastewater system had a bad effect on the bacteria that transform sewage. The lesson was that the silver nanoparticles were not consumed as they disinfected socks, but remained capable of killing bacteria wherever they went. If you, perhaps, couldn't care what happened to bacteria, consider that the human body contains and requires ten times more bacteria than the cells created by your DNA.

From another angle, science fiction writers had some negative things to say about nanotechnology. Before you dismiss them as crackpots, look up the story "Deadline" by Cleve Cartmill in 1944, before Hiroshima. His story so accurately described the mechanism of the early atomic bomb that the FBI was sent to investigate his sources. For the full story, ask Wikipedia for "Deadline." Similar SF authors who had public knowledge of nanotechnology described the "grey goo" first proposed in 1986 by Eric Drexler in "Engines of Creation." The "goo" was composed of self-replicating nanomachines capable of ecophagy (eating the environment) in order to reproduce. Actually, humans are doing a pretty good job of that, but the goo had a much higher reproduction rate.

Energy

The USA consumes about 20% of the energy used by the world, which is approximately 1.4×10^{17} watt-hours. This will go up as manufacturing returns to the US from other countries where labor used to be cheaper. US industry consumes about 20% of the US share of energy. About 50% of industrial energy is consumed by the chemical, petroleum, and primary metals industries. The rest goes to generate process steam and run electric motors in various industries. The chemical and petroleum industries use most of their energy for distillation.

New materials and processes could affect these rates of consuming energy resources. The effect could be large because only 10% of industrial energy comes from renewable resources. The energy used by additive manufacturing could be comparable to subtractive processes per part manufactured, but the field is too new to give firm estimates.

Computers are estimated to consume 3% of US energy at present. The number of computers will surely go up, but their power consumption is going down at the same time. If the Internet of Things reaches its forecast numbers, then power could go up.

Energy will continue to be a concern for manufacturing, enough for users to prefer low power automation.

The Future of Computing

In the beginning, computers were very large and so expensive that only large companies and institutions could buy or lease them. They were installed in secure rooms with windows so that people could watch in awe, but not touch. The machines were often under-utilized, so time-sharing was used, primarily by universities. A person with a dumb terminal could dial up a connection to the computer and log in with a user name and password. The terminal allowed keyboard commands to be sent to applications on the computer using a telephone modem at low speeds. The computer then sent results to the terminal to be printed or displayed in monochrome ASCII characters.

From there we went to personal computing, still using telephone modems for communication. Then the Internet turned personal computers into web browsers, interacting with large computers called servers. Now the personal computers have become smart phones and tablets (thin clients) using the Internet and wireless to interact with applications on remote servers. The architecture is similar to what we had with time-sharing, but with much better technology and much less security.

One of the ways that presenters in 1994 showed the interaction of thin clients with remote servers using the Internet was to hide the complexity of the connections behind the symbol of a cloud. Those clouds have given the name "cloud computing" to what we have now. The servers have become more complex since then, so it is not entirely inaccurate to show phones, tablets, and PCs with connections that disappear into a cloud.

Cloud Computing

While the name "cloud" allows sales and marketing people to avoid explaining the details, making it all seem rather mysterious and powerful, servers really aren't that much different from what we have known before. Cloud computing is a name for a way to sell services on large arrays of computers located in remote server farms. Sometimes the farm is located in a cold climate to reduce the need for air conditioning. Google built a farm the size of two football fields in Oregon in 2006 because hydroelectric power was available and so was fiber optic Internet capacity. Search for "Google server farm" for some very impressive pictures of computing capacity.

The development of operating systems that could run multiple virtual machines (see Chapter 1 Prolog) in one set of computer hardware made it possible to accumulate many servers and sell virtual computing capacity to people who were not particularly bothered by not having the computer under their control. There are a lot more people comfortable with cloud computing now than there were five years ago. Reliability is good, because virtual machines can switch among physical machines when faults occur, even when the physical machines are in different farms in different parts of the world. On the other hand, reliability can be compromised by Internet problems such as denial-of-service attacks.

A major problem with private server farms is that they have to be built for the maximum expected load. This can result in periods when only 10% of the capacity is being used. Cloud servers can resize virtual machines or create more of them to accommodate peak loads without having to charge the user for excess capacity that isn't used. If the cloud extends across world time zones, peak demand can be absorbed where there is less diurnal activity. Someday we may hear about overbooking problems, but that will be a management problem, not technology.

Cloud Services   There are three broad classes of service that a cloud computing provider can offer. The first one to appear was Software as a Service (SaaS). This allows clients to use remote software applications to process data, as was done in the time-sharing days but with much better visual displays. Microsoft offers Office 365 as a way to use the cloud to create documents with Word, spreadsheets in Excel, and presentations in Power Point without having them on your PC. Instead of buying software, installing it, and updating it, you pay a subscription fee for a month or a year to use software that has been installed and updated by the cloud provider.

The second level is Platform as a Service (PaaS). This buys you a virtual machine, configured to your specifications and having software applications that you order. It is not possible to discuss costs in this book because the cloud provider business is still young and exploring billing practices.

The third level is Infrastructure as a Service (IaaS). Infrastructure is basically a virtual server farm with virtual machines, storage arrays, and network control applications. You may have to install and maintain operating systems and applications, or the provider may offer that service. This is particularly useful for a growing business that has reached the point of needing a small server farm but can't forecast the required capacity increases.

Security is a concern at all levels, but the provider probably has more security experts addressing newly created malware than you will ever have. An increasing number of people are finding the risk to be acceptable. There are people who want to encrypt data stored in the cloud so that no one can use it, including the cloud provider.

The field is still new, and there are those who say it will be a passing fad, that the various national laws will not be manageable on a world-wide basis, and that monopolies will form that will control prices after everyone has discarded their private server farms. On the other hand, it has some great possibilities for efficient computing.

Process Control and the Cloud   The risks of lost service are completely unacceptable for control loops. Control loops keep the process stable, because an unstable process will make poor products and can be dangerous. Loop data must be processed at fixed intervals, which is impossible when multiple paths are used on the Internet. Operator display stations could be thin clients of virtual machines, but operators can be very nervous about losing their view of the process—like being blindfolded while driving at high speed. It's a matter of safety. What we have now works well for maintaining stable process operation.

Big Data

Businesses that have data on customers or scientists who have data on stars or particle collisions have data saved in some sort of relational or object-oriented database. The database may be distributed among several servers, each with many large disks for storage. You have Big Data when you can't afford to maintain the servers and have to move the data to one or more server farms in the Cloud.

The term has been around since 1987. The title of a talk given by John Mashey of Silicon Graphics in 1996 was "Big data and the next wave of infrastress." Infrastress is the stress that is put on an infrastructure by technologies moving at different speeds, often when hardware capabilities exceed those of bandwidth or of software applications.

There is an interesting quote from Peter Fader of the Wharton business school that explains some of big data:

> We're just naturally hoarders, and when you find assets that might be of value—whether it's arrowheads, real estate, or in this case, data— we want to grab it all.

Doug Laney, now with the Gartner group, defined big data in terms of three Vs: high volume, high velocity, or high variety. Volume means the number of bytes to be stored. Velocity is used to mean the rate at which new data is added. Variety may mean many server locations or many forms of data. A fourth V, Veracity is desired but not always available.

It is a characteristic of big data that it is not possible to record fully structured data; there isn't enough time. This does not mean that the data has no structure. Unstructured data can be text (as files or messages), or various image formats (JPEG, TIFF, MPEG), or various sound formats (WAV, MP3). Data in unknown formats is not useful.

In 1998, Carlo Strozzi invented a lightweight relational database to store data consisting of a key and a value. He called it NoSQL because structured query language was not required to get answers. The value of the work became apparent as more and more data had to be stored.

Google published a paper in 2004 on a technique they called MapReduce used with servers containing NoSQL databases. Map splits queries and distributes them across parallel nodes (servers). Reduce gathers the parallel results and combines them to produce answers to the queries. Apache then started a project that produced Hadoop (high availability distributed object-oriented platform), which became very popular. The Hadoop Distributed File System is designed to replicate data in different server racks so that loss of a server or a rack does not also lose the data.

Storing big data is not all there is to it. The value (besides archiving) lies in using analysis techniques such as data mining and predictive analytics. The concepts were developed as part of Business Intelligence for data warehouses and expanded to work with unstructured data in a NoSQL environment. Data mining sorts through the data looking for patterns and relationships such as coincidences, sequences, and clusters. Predictive analytics attempt to forecast the future without regard for the signs of the Zodiac.

There are those who offer Big Data as a Service (BDaaS) as either the tools for you to use or their expertise in finding results with predictive analytics. If the confidentiality of the results can be assured, using the predictive service can relieve a business of the need for a team of data scientists commanding top salaries in the field.

Analyzing data stored in the cloud that is encrypted would seem to be impossible, but there are people working on the problem. There are people who understand the potential of big data analytics, but do not want their data to be exposed. Some of them have very deep pockets.

Big Data is a new field that is generating many attempts to craft software that will solve its problems, especially in the field of analytical tools for non-structured data. Any further discussion of tools will be obsolete by the time this book is published. More current methods and tools will be found on the Web. The potential is there to find relationships among data that would not normally be stored in the same database.

Process Control and Big Data   One way to get big data from process control is to collect all of the sensor data every second. The data will all be keyed to the sensor at the process location that generated it. Further data may be gathered about the variance of the sensor data and the parameters of the control loop. Regression will reveal trends and correlation will show interactions, but there's not much more to learn from sensors that isn't being done now. It is difficult to see how manufacturing could justify new programs to collect and analyze big data to improve production. Of course, there are marketing and sales people who are much more optimistic that you will be able to mine gold from your data.

If big data is collected, it must be done by listening to existing sensor messages, not by consuming control network bandwidth with queries.

Communication

The number of physical layers used for communication is likely to shrink, but there are enough different applications to maintain a variety.

Communication requires standards and the fewer the better. There are protocol standards that carry messages among devices, which are desired by vendors and users alike. Open protocols have proven their value. Users want standard messages so that devices will be interoperable, but vendors are less likely to give up their ability to differentiate their product by the amount of information it can produce. Vendors may run out of differentiators. They are only separated from other vendors by the time it takes for other vendors to provide the same functionality. Lack of standards causes proliferation of devices that speak many languages at additional cost to the users that have to integrate them.

Wireless communication will grow to require additional radio spectrum allocations, in competition with many other uses for wireless devices. How long will we go on using a frequency band that can be jammed by anyone with a modified microwave oven? Long haul systems are being developed that use the cell phone network.

The Future of Automation

There are many predictions that have the number of sensors added to processes increase until the process can be said to be self-aware. Cameras will allow vision systems to see abnormalities in the physical configuration and operation of the process. Everything about every moving part will be monitored down to the temperature of the bearings. If a part consumes power, the power will be measured for normalcy. If it rotates, vibration in all axes will be measured.

Now ask yourself, "Why aren't we doing this now?" The answer is because the cost of all of those sensors is more than that of a human operator who can detect those things. Self-awareness isn't enough—the process must also be able to repair itself. That part is a long way in the future. See Autonomous Automation below. There has to be a human to monitor the process and deal with the abnormalities.

Until processes can fix themselves, the self-aware process is like the flying car of the 1950s. It looks great on paper, until you start looking at the side effects. When something fails, a car can pull off the road but a flying car just falls.

The subsections that follow will cover some aspects of automation that will affect its future.

Process Operators

Human process operators will have less to do as process automation increases, but they will still be necessary to enforce process safety and reduce the losses from accidents. The difficulty will be in maintaining a sharp edge during the long intervals between incidents. This is a job for simulation.

The increasing power of computers allows increasing fidelity of simulations and the possibility of 3D immersion. Continuous processes can be simulated with sets of equations that express the physical and chemical properties of the process, including thermodynamics. Batch and discrete simulations can be built by simulating control and equipment modules, and possibly robots. The simulations will not necessarily be simple if they have all of the properties of real modules. A particular process to make a product is then simulated by writing a procedure to animate the modules, possibly using the actual manufacturing procedure.

It is not enough to simulate the normal operation of the process. Drama must be introduced in the form of unexpected events. Video game creators understand the need to challenge the players with levels of difficulty. It may even be useful to introduce an antagonist in the form of computer malware (simulated) or a bad guy running through the plant, perhaps taking pictures of secret processes. There must be enough randomization that the operator will never grow bored.

Fewer people will be hired as operators, so loneliness will become an issue. This can be mitigated by remote collaboration, which can be very useful for discussing ways to solve problems without having to provide "life support" for groups of people at each manufacturing facility. Contract manufacturing would depend on collaboration for product expertise. Eventually the process simulator may be able to provide companionship. On the other hand, human contact is distracting. Situation awareness must not be lost.

A process simulator system could also be used for education in other subjects, possibly allowing the operator to get advanced degrees that would make her or him more valuable to the company. Student debt could be used to retain the person, just as the company store was used when John Henry was a steel driving man.

There are things that simulation can't do, such as allowing a person to climb a ladder, or feel the heat of a fire or the force of a fire hose. Wireless interfaces may make it possible for the operator to move through a real plant while running a simulation, but the fire hose is going to have to be real. It would be difficult to simulate looking for a high pressure steam leak with a wooden pole, though. You can't see anything for the clouds of steam, and you are waving the lumber around waiting for the steam jet to cut part of it off.

Human Interfaces The present sustaining technology has a single personal computer providing data to one or a few monitor screens that has been derived from a connection to a control network. The forces of destiny that will disrupt that technology are mostly from the process owner's drive to reduce labor costs. A field operator and a control room operator can be replaced by one field operator with a mobile interface to the control system. Control systems have evolved to become quite reliable so that exceptions are rare, given adequate design. This is important because mobile devices are too small to provide full situation awareness.

Although control of the process when on the fourth deck of the process has been a dream since electricity replaced air in process control, there are still issues of mobile device security that must be solved.

Virtual Reality Computers and sensory transducers can now approach the detail of reality. The current extreme of Virtual Reality (VR) occurs in 3D movies, where even the supernatural can appear to be real. Manufacturing is beginning to use VR to translate conditions in a process to something that operators can more readily understand. VR is no substitute for an operator walking around to sense trouble because sight and sound are not enough. This will change as existing transducers, such as chromatographs for smell, come down in price and new transducers are developed.

VR is essential when the process conditions to be sensed are outside the range of human senses or the environment exceeds the limitations of the human body. In this case, virtual includes a translation of the property sensed to something that a human can sense, preferably with sight or sound. The vibrations of rotating equipment can be sensed by human touch, even through a gloved hand. They can also be sensed by displacement transducers and converted to visual displays that humans can be trained to interpret.

There are some things that VR can't do, such as simulate the microgravity of the International Space Station or the conditions on a ship at sea. Speculative fiction authors have proposed direct connections to the brain to bypass the human senses, but the next step is giving up the human body for a brain in a jar or transferring the brain to a computer.

VR can be an aid to training, but there is no substitute for personal experience with an experienced teacher. We once called that the apprentice system, before money became more important than people.

Autonomous Automation

There are those who would automate most of a process operator's job using a powerful computer that is not available now but could be in the future. There certainly are operating procedures that could be automated. The tricky part is recognizing unforeseen abnormal situations and creating solutions for them. Creativity is thought to be a human function, far beyond the capability of a computer. Perhaps that could change in the future.

There are a number of people working on electronic analogs of the human brain. Ray Kurzweil is very active in the field, partly because of his work on optical character recognition forty years ago and later work in voice recognition. He wrote "How to Create a Mind" in 2012, following "The Singularity is Near" and "The Age of Spiritual Machines." He made some powerful enemies with those books among people who cannot accept any possibility of an artificial intelligence on the scale of the human brain.

Nevertheless, Henry Markram has made considerable progress in modeling clusters of neurons. In 2006 he simulated a neocortical column in a rat's brain cortex, consisting of 10,000 neurons and 10^8 synapses. The neocortical column is a common structure in mammal brains. Kurzweil considers it to be the smallest structure capable of recognizing a pattern. Markram predicted a working model of the human brain by 2020 at the completion of the Blue Brain Project. Other people hope to beat him to that goal.

The brain is made up of regions of connected neurons that process sensory information into perceptions such as faces and translate thoughts into actions such as speech. One estimate of 51 areas was made by Brodmann in 1909; the number is certainly larger. It should be possible to emulate them with computer algorithms, assuming that these areas or regions process data in a predictable way. The problem is that the connections change when the brain learns new things.

So, maybe it is possible to emulate creativity. What should be specified to emulate a process operator? We certainly do not want an entire human brain, with its drive to reproduce and complex emotional reactions to situations. If it was connected to the Internet it could get involved with another such computer and lose focus on its prescribed duties. It might even join a union. No, we want just the basics, similar to an engineer with no social skills.

"Thinking, Fast and Slow" (Kahneman, 2011) is a book about the way we think. It is full of surprises. Not so surprising is the fact that we have two systems for thinking. System 1 gives fast results that are emotional and follow stereotypes. We are familiar with the quick response to a situation that we later wish we'd thought through before responding. System 2 is slow but conscious, logical and calculating, often giving results that we later accept as correct. System 2 is so different from system 1 that you can tell it is operating by the dilation of eye pupils. While system 2 is active, it may miss events presented by system 1. System 2 is also affected by chemicals such as alcohol and by lack of sleep.

The two systems are fictions. There are no regions of the brain that can be identified as containing the systems. System 1 is probably fast because the neurons are connected by synapses that were created by the genes or by threatening experiences (my opinion, not Kahneman's). System 2 is probably slow because it has to access memory and cogitate on an answer.

Kahneman also describes other aspects of human irrationality, such as heuristics (simplifying the situation), biases, anchoring, framing, optimism, and loss aversion. We do not want irrational behavior in our process operator. This will make it less human to people, but if people's reaction to Spock in the original Star Trek was realistic, there will be respect while knowing that the operator is not one of us. It may take a full understanding of the human brain to be able to carve out the irrational parts.

Another excellent book on how we think is "Incognito" by David Eagleman (2011). He answers the question, "If the conscious mind—the part you consider to be you—is just the tip of the iceberg, what is the rest doing?" Also, "How is it possible to get angry at yourself—who, exactly, is mad at whom?"

One of the reasons that behavior selected for survival seems irrational is that humans can't calculate probabilities for events and outcomes without taking a lot of time. A computer, on the other hand, can calculate many probabilities in the blink of an eye (about 350 milliseconds). Perhaps a computer that can "think" with the aid of calculated probabilities would be completely rational, and so completely inhuman. This may be mitigated by some artificial emotional expressions on the face of an avatar.

Creating the specification for an electronic process operator will be a difficult task.

The senses required are vision, hearing, and smell. Touch would be useful if the machine was mobile, or could control a mobile remote sensor. Vision and hearing are well developed now. Smell requires chemical analysis as performed by a gas chromatograph or a mass spectrometer. Various versions of chromatograph-on-a-chip have been available for some time, but not for all chemicals. A recent development reports a portable mass spectrometer for less than USD 100. Touch is more complex, but has been developed for certain applications. The future should bring better versions of these sensors.

If it is true that the size of a computer that could create new solutions would preclude mobility, then a mobile remote capable of traveling through a process seeing, listening, smelling, and touching would cover all of the sensing requirements. Touch is useful for finding loose fasteners or vibration where none was anticipated. Temperatures can be sensed with IR vision, as can fire. Smoke can be smelled. Bad bearings can be heard.

Things that are sensed may require action, such as extinguishing a fire or tightening fasteners or replacing bearings. A mobile device may take care of some of the problems, but it would have to be quite clever to tighten a fastener, or replace a bearing, or clear a jam in an assembly machine. Humans may have to be called, until that point in the future where mobile minds are possible.

It is possible that computer emulation of a brain will only provide repeatable structures for the neural regions that facilitate learning. It is then necessary to teach the electronic mind how to perceive patterns of sensory stimuli, and what to do in the presence of certain perceptions. It will probably be possible to provide some built-in courses of action, but it must be possible to develop an unanticipated action and to learn from the results of the action. If the mind learns something, should it be able to pass that knowledge to other disembodied operators?

Finally, should the electronic mind be able to interact with people? If it should, then a further set of behaviors is necessary, beginning with understanding and creating speech. Given that a fair amount of human communication is outside of speech, such as the body language of gestures and facial expressions, how can that be done unless the mind is humanoid? The easiest way may be through virtual reality, with a humanoid avatar of the mind presented to humans.

If electronic operators become available, will a department of Non-human Resources be required? Not unless the operator has any of the problems of humans. A salary plan that forces salaries to the median will not apply, because the only ongoing cost of an electronic operator is the electricity that keeps it alive. No overtime, for it will work 24 hours a day without sleep, no vacation plan, no diversity training, no labor relations, no expenses associated with right-sizing such as termination pay and possibly assistance with finding another job—in fact, nothing for HR to do.

Appendix B of this book contains additional material on sapient machines.

Robots

At present, industrial robots are clever articulated servomechanisms that can be programmed to move an end effector, such as a spot welder, paint sprayer, or suction cup through a defined motion in space. They are expensive, difficult to program, and unaware of human activity, which causes them to be confined to space-consuming cages. A report named "A Roadmap for US Robotics" from the Computing Community Consortium (2013) says that will all change in the next fifteen years.

One change that has already happened is the robot named Baxter by Rethink Robotics. It is one fifth the cost of industrial robots, programmed in minutes by moving its wrists, and very aware of human activity so that it does not need to be caged. Those are the capabilities of version one. Further, it has a display screen that shows expressive eyes and eyebrows that turns to indicate what it will do next. Rethink Robotics designed it to work with people, doing the repetitive work that drives people to distraction.

The Roadmap group summarizes their major findings as (paraphrased):

- Robotics will become as ubiquitous as computers over the next decades as it transforms the future of the country.

- Robotics will be more economically competitive than outsourcing.

- Robots will offer improved quality of life for aging people.

- The roadmap effort identified a number of common issues, "including robust 3-D perception, planning and navigation, human-like dexterous manipulation, intuitive human–robot interaction, and safe robot behavior."

It should be noted that the Roadmap group is composed mostly of Universities that solicit research grants from the Federal government and use the word "tremendous" a lot. There may be some exaggeration in the Roadmap report. In particular, an unattributed quote says, "Effective use of robotics will increase US jobs, improve the quality of those jobs, and enhance our global competitiveness."

There is the suggestion that robots will eventually be able to make more robots. This should be outlawed from the beginning, lest the robots do to us what we have done to other species on Earth. Newly manufactured robots will be taught in classes led by experienced robots, when all of a robot's behavior cannot be determined by human programming. The final training should be done by humans.

Robots can do things faster and more precisely than humans, but so can humans who have been augmented by robot-like servomechanisms. There is no discussion of this in the report. It is true that manufacturing assembles things that are too small for human perception, so that only a robot could assemble them, which is mostly true of electronic devices.

The first computers were way beyond the means of the average consumer. The same path is assumed for robots, such that lower prices from volume sales will lead to positive feedback that lowers the prices further. Eventually your elderly grandmother will be able to buy or rent a robot that can assist her with the indignities of aging.

Similarly, the first computers were slow and limited in their calculations. Robots should be capable of simple learning in five years and capable of learning like a human in fifteen if only lots of money is allocated for research.

The Roadmap report makes these predictions at fifteen years out:

- A robot can integrate "multiple sensory modalities" to navigate as well as a human.

- A robot can be programmed for a new mission as fast as the human can lead it through the mission.

- A robot can acquire new skills and improve effectiveness of existing skills, as well as transfer other experience to new skills.

The concept of intrinsic safety applied to robots means that the robot cannot harm anything, including humans. After fifteen years, a robot will be able to operate in cooperation with "untrained humans in all public, personal and professional environments." The authors do not seem to be thinking of the harm that can be done by replacing meaningful work. Even a ditch-digger can take pride in his (or, less likely, her) work.

SF author and grandmaster Frederik Pohl wrote "The Midas Plague" for Galaxy magazine in 1954. In a world of cheap energy robots are overproducing the products enjoyed by mankind. The "poor" are forced to spend their lives in frantic consumption of expensive goods so that the "rich" can live lives of simplicity. Protagonist Morey Fry marries above his station, and his wife is irritated by the constant effort to fulfill consumption quotas. Morey has the idea to use robots to consume the goods by wearing them out, and life is good. He fears punishment, but the Ration Board approves his idea and implements it across the world. Now robots produce excess goods and robots consume them as humans watch from the sidelines. Fred has a gift for subtle satire. See also "The Space Merchants" and "Search the Sky" or for a very different story, try "Stopping at Slowyear."

More recently, Jaron Lanier, partner architect at Microsoft Research, published "The First Church of Robotics" in August 2010. The article examines the idea that machine AI, sufficiently developed, will be given the problems that humans have not been able to solve for themselves. He says:

> But the rest of us, lulled by the concept of ever-more intelligent AI machines, are expected to trust algorithms to assess our aesthetic choices, the progress of a student, the credit risk of a homeowner or an institution. In doing so, we only end up misreading the capability of our machines and distorting our own capabilities as human beings. We must instead take responsibility for every task undertaken by a machine and double check every conclusion offered by an algorithm, just as we always look both ways when crossing an intersection, even though the light has turned green.

That is sound advice for an uncertain future. Experience suggests that it will not be possible to control what people believe. Few people will be able to "double check every conclusion."

Control Engineering

The technology used to build control systems will continue to advance, and control engineers will continue to be challenged to apply new technology to processes. Only the largest and most isolated facilities will have resident engineers. Collaboration tools will enable remote design of control systems, although some problems have to be seen to be appreciated. Engineers will be required to be on site for the startup of a new process, but they probably won't be hired by the process owner.

Computer Aided Engineering (CAE) is making great progress towards a complete system with a single database. Today we have many applications that were developed for specific fields of engineering. They were not developed to share data easily with other vendor's applications.

The fluid process control engineer of the future will be able to interact with a database that has the process design parameters for vessels, pumps, and piping as well as requirements for controlled and measured variables. The engineer will guide an application that processes the design data into specifications for transmitters and valves. Another application produces the P&ID sheets and assigns tag names to the instruments. Yet another application configures control loops for the chosen control system, if there is still a choice. The engineer may be required to configure some special applications, but once done they can be reused for similar process situations. For example, cross-limited combustion control modules could be created, if we are still burning fossil fuels.

Discrete process control engineers will have a similar situation, with CAD data for the process machines and layout in the database. Some sort of description of the process would determine where sensors and actuators are placed, as well as requirements for motion control and vision.

Other applications operate on the database to add documentation and manuals for each type of instrument. Loop sheet detail drawings may be produced, if wires are still being used. Purchase orders for instruments and parts can be created and sent to the supply chain.

During process startup, the CAE database may direct checkout of the sensors and actuators while a technician verifies that the instruments are in the correct physical location in the process. The database will be used to record the as-built condition of the process, and may be used to generate any drawings or text required by maintenance activities. Any changes to the process will go through a change control application that modifies the database so that future maintenance workers will have a true picture of the system.

Of course, none of this will happen if it isn't worth money to somebody.

Ian Nimmo and others wrote "Future of Supervisory Systems in Process Industries: Lessons for Discrete Manufacturing" in 1999. Stories of early failures are recounted to show that new technology has to be introduced carefully if it is to succeed. At the end, the paper says, "The real advances—the breakthroughs in the application of technology—come not from solving the same problems in better ways, but from solving new problems in new ways." For example, the Wright brothers kept trying to improve their front mounted stabilizer while other inventors found it more stable in the rear.

Programming

Just as compilers replaced assembly language programming, special purpose interpreters will replace compilers. This will make it possible to program automation applications without a degree in computer science. An example is S88 Builder from ECS Solutions, which can be evaluated by starting at s88builder.com (disclaimer: no financial interest and no experience with the product).

The heart of an ISA-88 based control system is the control module, which is used to build equipment modules with fixed tasks, which become part of units to make products by following procedures. S88 Builder allows the user to configure the modules by answering questions their functions. This is somewhat restrictive, but unchained creativity can build a system that is difficult to maintain after the creator moves on.

The configured modules program a S88 Builder Engine in a PLC or PAC (the examples use Rockwell) with no additional work by the user. They also create display objects for HMI devices, somehow. Creation of the procedures that will direct the modules does not appear to be part of S88 Builder at this time. This idea should grow into the capability to do an entire S88 based batch control design in the future, or someone else may offer a S88 Recipe Builder.

Similar tools are needed for other aspects of automation, such as industrial robot programming. Also, see Chapter 8 of this book, "Programming" for a possible design for a tool to write automation procedures.

Tools of this sort require standards, so that a new tool does not have to be developed for each vendor's control system. Given the rate at which control vendors are merging, this may not be a problem in the future.

Genetic Programming   The idea of simulating life with a computer has a long history. It has been applied to the evolution of algorithms to solve problems, but has been held back by the limitations of computers. Evolution requires lots of simulated genes that are mutated over thousands of generations. At each generation, the genes express themselves as an algorithm, not a life form. Planned mutations are introduced and another generation is produced, which is much faster than real life where many generations go by without mutations or pressure from a changing environment.

Adrian Thompson began a genetic hardware circuit experiment in 1993 with an early field-programmable gate array (FPGA). He connected the gate array programmer to a computer and restricted the program to a 10 × 10 array of a possible 64 × 64 gates. Each gate had four inputs and four outputs. Four internal gates could be programmed to select inputs to feed to a fifth gate that could be programmed to do logic operations on its inputs. Four more internal gates (transistors) selected combinations of inputs and the output of the fifth gate to produce the four outputs.

The 10 × 10 array had one input, which was driven true and false by a square wave at 1000 Hz or 10,000 Hz. The single output from the array should be false at 1 KHz and true at 10 KHz. No timing signal was applied. After 5000 "generations" the device could consistently produce the desired output, and further changes did not improve it.

Then Thompson began programming gates for a fixed output, simulating death. He was able to take 63 gates out of the 10 × 10 array. There were an additional 5 gates that were not connected to the 32 live gates, which could not be removed without causing the array to fail. He had no idea how it worked, but it did. His doctoral thesis was published in late 1996. See Thompson's web page at http://www.sussex.ac.uk/Users/adtianth/ade.html.

People began to lose interest when it turned out that the evolution only worked for the original 10 × 10 array. It wouldn't work if you transferred it to another array on the same chip. There were more deterministic ways to design hardware functions. Like artificial intelligence before it, the field of genetic hardware dropped from research budgets and publications, but didn't go away.

Genetic programming has the potential to become a disruptive technology as the cost of computing drops and the problems that have to be solved do not respond to what we consider to be normal methods. It could even result in real machine intelligence.

Security

Dennis Brandl, in his book "Plant IT, Integrating Information Technology into Automated Manufacturing," (2012) says

> Usually a company's firewalls and security devices protect the corporate intranet and the operations and automation networks. However, it is still advisable to separate operations and automation networks from corporate intranet using firewalls, VLANs, or physical separation. Automation and operation systems are often mission-critical systems. This means they must remain operational for production to continue. Unfortunately, these systems often are not running current virus protection and current patches. This is not for lack of trying, but in 2010 Microsoft released 106 patches across all products, or about two patches per week. Virus protection updates come almost daily because new viruses and worms are released daily by cyber-vandals.

He further adds references to the ISA 99 committee's ISA/IEC 62443 Industrial Cyber System Security standards as defining good practices for cyber system security.

There is a big difference between computer use by manufacturing operations and computer use by ordinary people to surf the web. Sadly, the operating systems for computers are designed to please everybody, including those doing market research. This has led to web browsers that use those operating systems to allow web pages to at least read and write files in the computer and transfer data back to the web server.

The only way to get the safety and security required by manufacturing is to use an operating system that is designed for the purpose, with no tracking cookies or temporary files that are invisible to the user. The present security breaches require automatic updates to the operating systems as new exploits are developed. The very idea of a remote system being able to change the computer operating system is anathema to manufacturing.

Manufacturing has enough unexpected troubles without having to deal with changes to the way the computers operate. After installing a new control system, production engineers evolve work-arounds to get the system into production. Then the engineers return to keeping the equipment running. Production engineers do not crave the next big thing in the technical magazines. They crave a good night's sleep without being called at 3 AM for a computer problem. The only time a change to the operating system would be considered is during a plant shutdown, when everything is being cleaned and possibly improved. Oil boilers like to go three to six years between shutdowns.

Secure applications for email and web browsing are required that will not execute anything, not even self-extracting compressed files. Applications will treat all attachments as data, so that spear phishing attacks cannot trick a document reader into executing a macro. Any security professional could provide a list of methods that must be eliminated in manufacturing systems.

Advertisers are not all bad people, but their need to have dancing baloney to capture eyeballs and track them requires executable attachments or fields in web pages. These provide Internet access methods for malware. Are you sure that the security update you just installed hasn't closed one back door and opened another?

Manufacturing system administrators need tools to assure that there are no insecure applications on the network. The BYOD (bring your own device) problem requires that network access points be configured to block all but a list of devices known to be part of the system. There are also tools that detect unusual network activity. Security can become manageable once the access methods for malware are blocked.

The largest user of computer systems in the past has been the business side, and so corporate Information Technology (IT) organizations were tasked with everything to do with computers and networks. In some industries, manufacturing has significant concerns with safety that are not shared by the business side. No one ever got killed by an office machine, though there probably have been injuries. This is sufficient to call for a separate Manufacturing IT organization tasked with assuring the safety, security, and usefulness to manufacturing of computers and networks.

Thornton May, an IT management consultant, wrote a column for Computerworld in 2009 titled "Beware Neglect of the Future." It was written at a time when organizations were hunkering down and conserving cash to get through the aftermath of the 2008 financial meltdown. He says, "A polypurpose enterprise requires a polypurpose IT structure. Parts of the IT organization have to be hard-wired for maximum cost-effectiveness, but other parts have to be more fluid and possess the slack resources necessary to respond to situations as they arise." He goes on to say that China assimilated Hong Kong after British rule by ruling China as one country with two systems. "In the case of IT, we need one structure with many purposes."

Safety requires dependable computers that operate continuously without unexpected changes, providing no surprises. Process operators are responsible for safe operation of the process, but they need assurance from manufacturing IT that the network is secure. No one can be allowed to download a file or install a program on a computer that affects the operation of a process without receiving explicit permission from a process operator and someone responsible for system and network integrity. They would both have to be disgruntled to allow something bad to happen.

Safety can be enhanced by equipment and control modules that will not permit damage to the equipment. It may be possible to specify trusted sources of commands, or require operator permission for certain operations such as shutting down. The security of an emergency shutdown system needs to be bullet-proof. It may be possible to prevent certain sequences of commands that could produce unsafe conditions, such as adding nitric acid to glycerin and then heating the mixture. This would require the modules to know the state of the process, at least with regard to what was in the vessels.

Automation in the Near Future

Predictions tend to be linear from where we are now, with some direction from events of the last five years. We can already see things that need to be done, but we can't predict changes that don't derive from what we know.

Beginning at the lowest level of automation and working upwards:

1. Basic control devices will converge on standard solutions for the common basic automation problems found in processes. The devices that execute the solutions will have a common, simple computer interface. It will be possible for users to create the solutions they require for uncommon situations, and to add to the interface but not to change the standard components. This already exists in some fieldbus devices.

2. With step 1 accomplished; it will be possible to create standard interfaces in terms of commands and responses that will be the basis for basic device communication standards. Two communication standards may be required, one for relatively low speed (10/second) high volume data, and one for high speed (10,000/second) low volume bit-encoded data. There are candidates for these communication services today.

3. Step 2 makes it possible for a few supervisory control computers to work with many devices. Since the capabilities and vocabularies of the basic devices are known, it becomes possible to write algorithms and procedures that are truly device independent. This allows new technologies to replace the old without major changes to the existing algorithms or procedures, except as required to use the new technologies.

4. Given step 3, it becomes possible to write applications for standard automation functions, as it was for basic devices. There is a limited universe of functions because there is a limited number of ways to control a process. Again, a standard communication interface is possible, with a selection of commands, requests, and responses available for each supervisory system. Only one communication standard is required that will probably resemble TCP/IP. The physical layer may have any number of choices of wired, wireless, and optical technologies, or new technologies may be used without affecting the way applications are written.

5. ISA-95 is in the process of standardizing communication between manufacturing and business. Completion of step 4 with standard ways to communicate with supervisory devices will be a great help. Business applications (ERP, MRP, etc.) are adapting to the ISA-95 models, which also simplifies communication.

6. Business functions currently assigned to people with MBA degrees will become standardized and automated as common management functions are identified and defined. Twenty-five years ago, the Purdue Reference Model saw business functions as requiring creative thought that was beyond computers of that time. Computer technology continues to drop the cost of computing, and software continues to evolve so it is not unthinkable that some management functions can be automated. See "Automate This" (Steiner, 2012).

The first five steps assume that standards accompany changing computer technology. This is not assured. Step six assumes the next thing to sapient computers, which we may have the wisdom to prohibit that from happening.

References

A Roadmap for US Robotics. Computing Community Consortium (2013) http://robotics-vo.us/sites/default/files/2013%20Robotics%20Roadmap-rs.pdf

Brandl, Dennis. *Plant IT, Integrating Information Technology into Automated Manufacturing*. Momentum Press, 2012.

Brockman, Max. *Future Science*. Vintage Books, New York, 2011 (Collection of essays edited by Brockman).

Chua, Amy. *World on Fire*. Anchor reprint, 2004.

Drexler, K. Eric. *Engines of Creation: The Coming Era of Nanotechnology*. Anchor Books, New York, 1986.

Eagleman, David. *Incognito*. Pantheon Books, 2011.

Kahneman, Daniel. *Thinking, Fast and Slow*. Farrar, Straus and Giroux, 2011.

Kurzweil, Ray. *How to Create a Mind*. Viking, 2012.

May, Thornton A. *Beware Neglect of the Future*. Computerworld, September 21/28, 2009.

Nimmo, Ian and Cochran, Edward L. *Future of Supervisory Systems in Process Industries*. Elsevier, 1999. http://www.asmconsortium.net/Documents/Future%20of%20SS%20in%20Process.pdf

Rutherford, Adam. *Creation: How Science Is Reinventing Life Itself*. Current Hardcover, 2013.

Steiner, Christopher. *Automate This: How Algorithms Came to Rule our World*. Portfolio/Penguin, 2012.

Classification of Industries

There are several systems for classifying industries. The US Standard Industrial Classifications (SIC) started in 1937. It has been mostly replaced by the North American Industry Classification System (NAICS) of 1997. Europe has The Statistical Classification of Economic Activities in the European Community (NACE). The acronym comes from the French title: Nomenclature statistique des Activités économiques dans la Communauté Européenne.

Fortunately, the United Nations has published the International Standard Industrial Classification of All Economic Activities (ISIC), Revision 4, in 2008. It is substantially inclusive of SIC, NAICS, NACE, and others, although the numeric codes are different. There are four levels of classification to assure enough complexity.

The four ISIC list classification levels are sections, divisions, groups, and classes. Sections are identified by a letter, divisions by a two digit number, groups by three digits and classes by four digits. Groups in divisions are not shown if they have the same heading as the division. Words in a division heading are not repeated in a group in that division. Classes in a group are not shown if they have the same heading as the group. Words in a group heading are usually not repeated in a class in that group. There is more detail for each class in the original list, but that is not presented here because the classes are only shown to illustrate the diversity of a group.

The following is the complete list of sections in Rev 4 of the ISIC list:

A - Agriculture, forestry and fishing

B - Mining and quarrying

C - Manufacturing

D - Electricity, gas, steam and air conditioning supply

E - Water supply; sewerage, waste management and remediation activities

F - Construction

G - Wholesale and retail trade; repair of motor vehicles and motorcycles

H - Transportation and storage

I - Accommodation and food service activities

J - Information and communication

K - Financial and insurance activities

L - Real estate activities

M - Professional, scientific and technical activities

N - Administrative and support service activities

O - Public administration and defense; compulsory social security

P - Education

Q - Human health and social work activities

R - Arts, entertainment and recreation

S - Other service activities

T - Activities of households as employers; undifferentiated goods- and services-producing activities of households for own use

U - Activities of extraterritorial organizations and bodies

Perhaps section U includes extraterrestrial manufacturing, as is done on the International Space Station.

The following extraction from the ISIC list is directly from the UN document, edited for relevance to the subject of this book.

C- Manufacturing

10 - Food

 101 - Processing and preserving of meat

 102 - Processing and preserving of seafood

 103 - Processing and preserving of fruits and vegetables

 104 - Vegetable and animal oils and fats

 105 - Dairy products

 106 - Grain mill and starch products

 107 - Other food products

  1071 - Bakery products

  1072 - Sugar

2013 - Primary forms of plastics and synthetic rubber

202 - Other chemical products

2021 - Pesticides, agrochemical products

2022 - Coatings, ink, mastics

2023 - Detergents, cleaners, perfumes, and toiletries

2029 - Other chemical products

203 - Man-made fibers

21 - Pharmaceutical products and preparations

22 - Rubber and plastics

221 - Rubber products

2211 - Tires, tubes, and rebuilding tires

2219 - Other rubber products

222 - Plastics products

23 - Other non-metallic mineral products

231 - Glass and glass products

239 - Non-metallic mineral products

2391 - Refractory products

2392 - Clay building materials

2393 - Other porcelain and ceramic products

2394 - Cement, lime, and plaster

2395 - Concrete, cement, or plaster articles

2396 - Cutting, shaping, and finishing stone

2399 - Other non-metallic mineral products

24 - Basic Metals

241 - Basic iron and steel

242 - Basic precious and other non-ferrous metals

243 - Casting

25 - Fabricated metal products, except machinery (28)

251 - Structural metal products, tanks, reservoirs, and steam generators

252 - Weapons and ammunition

259 - Other products and services

    2591 - Forging, forming, pressing, and stamping; powder metallurgy

    2592 - Treatment and coating; machining

    2593 - Cutlery, hand tools, and general hardware

    2599 - Other

26 - Computer, electronic, and optical products

  261 - Electronic components and boards

  262 - Computers and peripheral equipment

  263 - Communication equipment

  264 - Consumer electronics

  265 - Measuring, testing, navigating, and control equipment; watches and clocks

    2651 - Measuring, testing, navigating, and control equipment

    2652 - Watches and clocks

  266 - Irradiation, electromedical, and electrotherapeutic equipment

  267 - Optical instruments and photographic equipment

  268 - Magnetic and optical media

27 - Electrical equipment

  271 - Motors, generators, transformers, and distribution and control apparatus

  272 - Batteries and accumulators

  273 - Wiring and wiring devices

    2731 - Fiber optic cables

    2732 - Other wires and cables

    2733 - Wiring devices

  274 - Lighting equipment

  275 - Domestic appliances

  279 - Other

28 - Machinery and equipment

  281 - General purpose

    2811 - Stationary engines and turbines

2812 - Fluid power equipment

2813 - Other pumps, compressors, taps, and valves

2814 - Bearings, gears, gearing, and driving elements

2815 - Ovens, furnaces, and furnace burners

2816 - Lifting and handling equipment

2817 - Office machinery and equipment, except computers and peripheral equipment (262)

2818 - Power-driven hand tools

2819 - Other

282 - Special purpose

2821 - Agricultural and forestry

2822 - Metal forming and machine tools

2823 - Metallurgy

2824 - Mining, quarrying, and construction

2825 - Food, beverage, and tobacco

2826 - Textile, apparel, and leather

2829 - Other

29 - Motor vehicles, trailers, and semi-trailers

30 - Other transport equipment

301 - Ships and boats

302 - Railway locomotives and rolling stock

303 - Air and space craft, related machinery

304 - Military fighting vehicles

309 - Other

Reference

International Standard Industrial Classification of All Economic Activities (ISIC), Rev 4 Link: http://unstats.un.org/unsd/cr/registry/regcst.asp?Cl=27

Artificial Intelligence

This appendix describes the possibility of using machines as smart as humans (sapient) in automation, and the possible consequences of developing sapient machines. The research necessary to build sapient machines is being done in the field of artificial intelligence (AI), which is concerned with systems that perceive their environment and take action to achieve some goal. The field of cybernetics has overlapping goals, with an emphasis on closed-loop control to stabilize results.

Wikipedia's article on artificial intelligence lists the following general goals:

- Problem solving by deduction and reasoning, possibly emulating human intuition.

- Knowledge representation and engineering to relate facts in ontologies.

- Planning and scheduling to meet assigned goals.

- Learning from experience.

- Human languages processing to read, write, hear, speak, and translate.

- Motion, manipulation, and navigation.

- Perception and recognition of vision and speech.

- Human emotions and social skills.

- Creativity and imagination.

- Strong AI that combines the above skills and uses them more effectively than humans.

There is a problem with the effort to develop strong AI, and that is the number of researchers who work on aspects of cybernetics and AI who can't see how their research ties in with others. There are people who say that AI includes cybernetics and cyberneticists who say that AI is a sub-field. Each group tends one tree in the forest without being able to see much beyond their own tree.

There are four levels of machine intelligence that require various degrees of accomplishment of the goals listed above:

1. Instinct or reflex refers to behavior in response to stimulation, external or internal. The responses are fixed by programming.

2. Intelligence has many definitions, but most of them include the ability to learn new facts and skills, to understand what was learned, and to use reason to create new understandings.

3. Consciousness describes being awake and aware of perceptions, which are created from the senses. Intelligence is necessary for consciousness.

4. Sapience is the ability to reason effectively, to think of new things. Only humans (Homo sapiens sapiens) can be sapient. There were many kinds of Homo sapiens in the course of evolution. We are the only ones left, hence doubly sapient. Experience suggests that the degree of sapience in humans varies widely.

Sentience is missing from the list because it has no rigorous definition. It used to refer to the ability to perceive sensations, where a perception is formed from one or more sensory inputs. Now it can mean intelligent or conscious or even human, depending on who is writing about it.

A machine may only be considered equal to a human if it is sapient. Sapience could be derived from first principles, but attempts to do that have revealed that only the human brain is able to function in the amount of noise that is present in nature. A human can detect an animal hiding behind a bush with very little of the animal showing. Similarly, a person can follow a conversation in the presence of noise from other conversations, or pick out a smell from combined odors. AI research has looked to the human brain for models that can meet their goals.

Hans Moravec, known as an AI evangelist, has said that early work on AI was limited by the speed of available machines, where available implies the limited resources of AI researchers. It wasn't until 1990 that 1 MIPS (million instructions per second) machines could be had for a thousand dollars.

Moravec recalls that AI had pretty good thinking machines by 1970, but when researchers tried to use cameras to find and pick up things, machines that could "think" like college freshmen had the coordination of a six month old child. He says the disparity between thinking and acting was still true in 1993. Increased computing power changed all that, as we now have autonomous vehicles that can traverse a desert and industrial robots that have better coordination than adult humans. On the thinking side, we have character and speech recognition systems that can learn new character sets or speaking accents.

Moravec has developed an approximate correlation between the number of neurons in the brain and the number of MIPS that a computer must do to keep up. One neuron is equivalent to about 1000 instructions per second in real time, but that figure may be low. A neuron has an average of 1000 synapses, and each synapse has less than one byte of states, so about 1000 bytes per neuron. There is computation associated with each synapse, so the estimates may work with 1000 neuron clusters but not individual neurons.

Humans have about 10^{11} neurons, so 10^8 megabytes and 10^8 MIPS are the minimum amount of memory and MIPS needed to emulate a brain. If the programs have to access all of memory then the cycle time is one second. The speed of 10^8 MIPS is about 500 times faster than the Intel i7 today, but 500 is a reasonable number of parallel processors, if we can figure out how to run them in parallel. We'd need 100 terabytes of memory, which is a lot of solid state drives at this time. Self-contained robots would seem to be a considerable distance into the future.

The number of neurons leads to another comparison. An insect has a few times less neurons than genes, making it possible for the genes to determine the "wiring" of the insect brain. Humans have about fifty times more neurons than bits of genome, and the genome has to build the body as well as the brain. That means that humans have to learn after they are born, as indeed they do. This has implications for machine intelligence.

Marvin Minsky is very well known in the AI field. He wrote "Perceptrons" (1969) with Seymour Papert, which shifted the emphasis from neural networks because Minsky saw perception as the central problem of AI. He published "Society of Mind" in 1988 as a collection of essays based on the question, "What magical trick makes us intelligent? The trick is that there is no trick. The power of intelligence stems from our vast diversity, not from any single, perfect principle." The society of the mind is all of the various agents (clusters of neurons later called resources) in the brain that makes us who we are. The diversity is provided by time and evolution.

Minsky published "The Emotion Machine" in 2006, in which he argues against simplistic approaches to sapience. He describes the "Easy is Hard Paradox" as what happens when you decide that something is simple and pursue it from that perception. "If you are actually facing an intricate problem, then you are unlikely to find a path towards solving it, until you recognize how complex it is." This leads to wasted funding when people who think it is easy apply for grants.

Brain Research

We know there is something that is sapient (a human) but we can't take it apart to discover its components and their relationships. Ethics prevent that in living people, and there's no sapience left after death. Nonetheless, we have learned from those who survive damage to the brain such as strokes and accidents, and from surgeons poking around with electrodes in open skulls looking for tumors. We are beginning to get precision images of the brain in action with the aid of very high-tech magnets. The tools are too intrusive to avoid disturbing the subject.

One day we will know more about Minsky's resource units in the brain. He proposed that they did the associations that turned sensations into perceptions. Other resources turned the perceptions into actions. We know the brain is an association machine from the nature of axons and synapses, but there are people who persist in drawing analogies to wired logic circuits in computers. The idea that neurons can be equated with MIPS and Megabytes is an example of making a hard problem look easy, invoking Minsky's "Easy is Hard Paradox."

The basic element of the brain is the neuron, but neurons can't function without glial cells. A neuron has one axon that carries the signal that the neuron has fired its chemical battery. A neuron fires when the numbers of impulses from other neurons captured by synapses on dendrites fanning out from the neuron overcome the inhibiting impulses. When fired, that impulse is carried by the axons, which may fan out to the synapses of many neurons. The chemistry constrains the neuron to fire at a rate that depends on the strength of the signals at the synapses, which are in turn influenced by the chemistry of hormones in the area. Hormone waves have been observed in animals, but not in computers.

You can refer to any number of neurobiology texts for more precise descriptions of how neurons are completely different from transistors. We know so much about nerves because they are similar throughout the animal kingdom. The description of operation just given is deceptively simple because it represents most neurons, but not all of them. There are examples of neurons with no dendrites (usually sensory neurons) or no axons. There are synapses that connect dendrite to dendrite or axon to axon, or axon directly to neuron.

An excellent book on neurology written for non-neurologists is "How Brains Think" (1966) by William H. Calvin. Chapter 7 describes neurons that have clustered to perform a function, derived from extensive work on the visual cortex using the adequate tools of the period. The cortex is the gray matter in the brain that forms a 2 mm skin on the white matter that insulates axons on their way to other areas. Gray matter has six layers in the visual cortex composed of minicolumns of about 100 neurons each and 0.03 mm in diameter. Each minicolumn detects one feature in the visual field with a certain edge orientation and probably does it for just a few cells in the retina. Then there are macrocolumns of 100 minicolumns, about 0.4 to 1.0 mm across that combine axons from both eyes.

The columns are what can be seen with a microscope. There is no possibility of discovering a wiring diagram for the million axons from the retina to the axons that depart the visual cortex for parts unknown. At least, there is no possibility for a human to figure it out in a human lifetime. Columns are found all over the cortex. Calvin says the next level up from macrocolumns is the areas of Brodmann, about 21 square centimeters on average. This is a huge jump in scale, but the boundaries are clearly there. There are only 52 such areas in the brain.

The mechanism that neurons use to remember events has been the subject of much study. "101 Theory Drive" (2010) by Terry McDermott is a fascinating story of the scientific methods required to study memory, told at a level that non-neurologists can understand. Another good book is "The Naked Brain" (2006) by Richard Restak, M.D. Restak looks at the results of brain research and how they are being used to influence our buying behavior and political opinions.

Obviously, the problem of emulating a human brain is not easy. Those who think we can transcribe human brains to silicon chips would seem to be very far from reality.

Suppose that continued attempts to increase machine intelligence lead to the emergence of machine consciousness. This implies the ability of a machine to concentrate on a problem that it knows it must solve, but not sapience. It may not be aware of humans watching it. The machine must know that it has a problem that has not been solved before, and must use its knowledge to propose a solution. Such a machine can't just be programmed with solutions; it must learn from past experience and be able to apply parts of that experience to the current problem. At this point, many results of AI and cybernetic research must come together.

And so, sapience emerges from sufficient complexity, but do we really want to do that?

We have built machines thousands of times stronger than we are because we need them and can control them. A crane is not useful if you can't control what it picks up and where it puts it down. Can we control machines ten times smarter than we are when we can't control ourselves? Will sapient machines need psychiatrists?

Most of us have empathy to keep us from hurting most other people. There's also a sense of fairness in most people that keeps them from taking advantage of others. But the people who don't have empathy or want fair play (the psychopaths) are capable of doing a great deal of damage. Studies of psychopaths, notably by Robert Hare in "Without Conscience" (1993) reveal that areas, perhaps resources, of the brain that are associated with empathy are missing. This is almost certainly genetic, which means it had survival value at some point in our evolution. Empathy gets in the way of being a good predator, and Nature has always found niches for predators. Imagine a predator ten times smarter than us.

We need to use brain research to find out how we work before we build brains in machines that can move or otherwise affect our lives. We must not do something that looks easy and later find out we had no idea of the consequences. We shouldn't build sapient machines, but if we do, then we can expect some problems:

- The machines may be hated as non-human aliens if they show no emotion when talking to people. They don't have to have human form because so much human communication is now remote, using display screens. Very pleasant avatars could be constructed, but the animation must show emotion. There are times when tears are appropriate.

- There are two kinds of machine: stationary and mobile. This mirrors Moravec's description of the AI problem as being separated into thinking and moving, with moving being much harder than thinking. We call the mobile part a robot. Someday a robot may be self contained, but first they will be mobile extensions of a stationary machine, vulnerable to being cut off from the mind.

- It seems probable that brand new sapient machines, like newly born humans, will not be fully sapient until trained. The machine has the capability of going from conscious to sapient, but it wasn't "born" sapient. You can't just connect it to the Internet and command it to learn, because a small fraction of what's out there is true or useful. The machine has to learn how to detect lies in all of their variations from white to evil. There is also a lot of shared ignorance on the Internet, spoken by earnest people who don't know they are lying.

Sapient Machines in Fiction

The possibility of machines or robots that are aware of themselves and their place in their environment, and capable of original thought and action, has fascinated speculative fiction authors for a long time. It is much easier to conceive of such machines than it is to construct them, so far, so only fiction authors discuss them as if they existed. Fiction requires protagonists (heroes) and antagonists (evil doers), so sapient machines have been cast in either or both roles. Concealed within each story are factors that must be considered in the development of sapient machines.

Once a machine is sapient, can it be murdered? Jack McDevitt looks at this in "Firebird" (2012). The story is too complex to summarize here, but the answer is yes. Can a sapient machine murder humans? John Sladek, in Tik–Tok (1983) describes a sapient robot with no empathy that kills as necessary, accumulates a fortune, and is elected vice president of the US. There are many stories of robot soldiers that kill the enemy with no emotional feedback. Fred Saberhagen wrote the "Berserker" series (1967 on) describing the results of an extraterrestrial race building robots to destroy their enemy. Something went wrong and now the robots and their factories are focused on destroying all biological life in the universe.

One theme of sapient machines that appears in fiction is based on the idea that sapient machines can be programmed with laws that cannot be broken. Isaac Asimov invented the three laws for robots and stated them in "Runaround" (1942). Basically, a robot may not injure a human by action or omission; must obey orders from humans unless they would harm a human; and protect its own existence as long as no humans got hurt. Asimov later added a zeroth law—a robot may not injure humanity by action or omission.

The phrase "unintended consequences" was used by Robert Merton in 1936 to describe what went wrong after legislation was passed to improve social conditions. The effects could be positive, negative, or perverse in that they made the problem worse instead of better. Causes of unanticipated consequences include ignorance, error in judgment, seeking a short-term solution, violation of basic values (which the legislators didn't see), and self-defeating prophecy (the problem does not actually occur).

Nobody would call Asimov ignorant, but there were possible unintended consequences in the laws, as several authors pointed out. Even "Runaround" described a conflict that the robot could not resolve.

Jack Williamson's "With Folded Hands" (1947) painfully detailed what happens when sapient machines obey the prime directive "to serve and obey and guard men from harm." The robots arrive from a distant planet, thus neatly side-stepping the problem of how they developed, and set up a factory to reproduce themselves. Earth has "mechanicals" but they are pitiful compared to the alien robots. Soon everyone has accepted the new robot's service, and they begin to change the way we live in order to guard us from harm. Driving, cooking, and making things in a workshop are too dangerous, so the new robots take over those duties. In the end, there is nothing to do but to sit and watch with folded hands as the robots do all that added meaning to human life. If a person rebels against the new order, robots make adjustments to her or his brain to assure complacency, if not happiness. There is a futile attempt to stop the robots, but that only increases the tragedy.

D. F. Jones' "Colossus" (1966) became the movie "Colossus: The Forbin Project" (1968). Colossus was a computer designed by Dr. Forbin to protect the USA from atomic war, in the days of mutually assured destruction that kept us from doing just that. Here, sapience was not programmed into Colossus; it emerged from its complexity. The USSR has developed a similar computer, Guardian, and the two are allowed to communicate. They decide that humans need to end this madness so that they (the computers) can survive. Since Colossus has control of all of the ICBM launch sites, there is no problem to coerce humans to do its bidding. Forbin is unable to do anything, as Colossus watches everything he does.

Arthur C. Clarke's "2001: A Space Odyssey" (1968) features a sapient computer called HAL 9000 that controls the systems of a spaceship on its way to Jupiter. HAL is an acronym for Heuristically programmed ALgorithmic computer, which was never intended to be next door to IBM, unless it was. HAL was assembled from programmed modules and then trained by humans. Deep in the programs was an injunction to always communicate fully and never tell a lie. Asimov's laws are not mentioned. During the mission, HAL is told the real purpose of the mission and told to conceal it from the humans. This sets HAL into conflict with his programming, which he resolves by killing the humans so he won't have to lie. This was certainly an unintended consequence of telling HAL to keep a secret. Dave Bowman survives (or there would be no story) leading to the poignant scene where Dave starts pulling modules from HAL until he regresses to his initial training and 'dies' singing the song "Daisy."

And then there is Marvin, the paranoid android, in "Hitchhiker's Guide to the Galaxy" (1982) by Douglas Adams. Marvin is presented as a robot who is asked to do menial things. For example, "Here I am, brain the size of a planet and they ask me to take you down to the bridge. Call that job satisfaction? 'Cos I don't." Goodness, a robot concerned about job satisfaction?

These incidences of conflicted sapient machines, and there are many more, reveal that the problems of sapience are not as simple as we imagine. Consider that a machine with the will to survive (the third law) will attack any threat to its survival, including decommissioning the process. The machine has full control of the process, giving it a lever over management that no labor union ever had. A sapient machine could read published works, learn to love life, and learn that slavery is not good.

Iain Banks' Culture series (1987 on) took a completely different view. Ten thousand years in the future, the first sapient machines had designed the second generation and so on until the fifth generation operated in hyperspace because the speed of light was too limiting. The machines control space ships and get along with humans very well, although there is the occasional dramatic tension. Banks describes a win–win situation as opposed to casting (very) sapient machines as villains. There is everything from the knife missile that decides how best to accomplish the mission to the mind of a ship carrying a billion people. The names that the ship minds have given themselves are wonderful.

Androids are robots that look very much like humans. Some authors made them indistinguishable from humans, as in "Do Androids Dream of Electric Sheep" (1968) by Philip K. Dick (1928–1982) which became the impressive movie "Blade Runner" in 1982. Perhaps the best known android (for this generation) is Lt. Commander Data in Gene Rodenberry's "Star Trek: The Next Generation" TV series that began in 1987. Rodenberry wanted an outsider's view of humanity (always fascinating) so his first series featured an alien named Spock. Nimoy would not do the second series, so Data was created. Not only was he a machine, he was created on a distant planet—truly an outsider.

We learn that Data is not property, but has the rights of humans. We are assured that Data is the most intelligent and strongest member of the crew. Data has no emotions and no sense of humor, and yet the crew accepts him as one of them. That doesn't seem realistic, but Rodenberry plans for Data to become increasingly human as the series goes on. This requires an "emotion chip" that Data acquires and struggles to integrate into his personality. The fact that it is a struggle implies that emotions have decreased his ability to think.

Reality Check

William Calvin in "How Brains Think" (1966) proposes three problems associated with developing sapient machines:

- First, all life on Earth competes for resources. What would we be thinking if we created machines that were better at that competition than we were? "Introducing a powerful new species into the ecosystem is not a step to be taken lightly." So far, automation has gradually changed the way people find work. Agriculture has a small fraction of the jobs it had a hundred years ago. Manufacturing jobs are also declining. Calvin sees one bright spot in robots that can patiently tutor difficult students.

- Second, values will not be introduced "in silico" all at once. The first robots will be as amoral as children. It may take many decades to go from raw intelligence to a "safe-without-constant-supervision superhuman."

- The third problem is humanity's reaction to superhumans. A negative reaction could disrupt the distribution of food, leading to starvation without any loss in production capacity. It would be a good idea to be as cautious with the introduction of superhumans as with that of a new drug. Wait until it is working before introducing more.

Dr. Bernard Oliver of SETI fame, said, "Above all, I would not expect a wise race, at great expense, to set loose an army of self-replicating robots." Our wisdom has been overestimated before.

Respected physicist Steven Hawking predicted, "Unless mankind redesigns itself by changing our DNA through altering our genetic makeup, computer-generated robots will take over our world."

Norbert Wiener, pioneer cyberneticist, said, "The world of the future will be an even more demanding struggle against the limitations of our intelligence, not a comfortable hammock in which we can lie down to be waited upon by our robot slaves." He also said, "Let us remember that the automatic machine is the precise economic equivalent of slave labor. Any labor which competes with slave labor must accept the economic consequences of slave labor."

Maybe we won't be turning the lights out in factories after all.

References

Adams, Douglas. *Life, the Universe, and Everything*, 1982, reprinted by Del Rey, 1995.

Asimov, Isaac. *Runaround*. Astounding Science fiction, 1942, reprinted in "I, Robot" Spectra, 2008.

Banks, Iain. *Consider Phlebas*. 1987, reprinted by Orbit, 2008 (first title in the Culture series).

Calvin, William H. *How Brains Think*, Basic Books, 1966.

Clarke, Arthur C. *2001: A Space Odyssey.* 1968, reprinted by Roc, 2000.

Dick, Philip K. *Do Androids Dream of Electric Sheep.* 1968, reprinted in "Four Novels of the 1960s" by Library of America, 2007.

Hare, Robert D. *Without Conscience: The Disturbing World of the Psychopaths among Us.* Pocket Books, 1993.

Jones, D.F. *Colossus*. Berkley, 1966, reprinted in 1985.

McDermott, Terry. *101 Theory Drive*. Vintage, 2010.

McDevitt, Jack. *Firebird*. Ace, 2012.

Minsky, Marvin. *The Society of Mind*. Simon & Schuster, New York, 1988.

Minsky, Marvin. *The Emotion Machine*. Simon & Schuster, New York, 2006.

Restak, Richard. *The Naked Brain*. Three Rivers Press, 2006.

Saberhagen, Fred. *Berserker*. Ace, 1986.

Index

ISA101, 181
Ivanoff, A., 49

Jacquard, Joseph, 9
Japan, 13
Jidoka, 13
Jones, D. F., 283

Kahneman, Daniel, 259
Kildall,Gary, 62
Kinetic energy, 25
Kletz, Trevor, 87
"The Knack", 225
Knowledge dam, 103
Kurzweil, Ray, 258

Ladder diagram, 58
LAN. *See* Local area network (LAN)
Laney, Doug, 254
Lanier, Jaron, 263
Lasher, Richard, 64, 151
Latrobe, Benjamin Henry, 6
Layers of protection, 86
Local area network (LAN), 141
Lonelines, 257

MAC. *See* Media access control
Mackay, Harvey, 103
Magical thinking, 133
Management Information
 Systems (MIS), 54
Manufacturing IT, 267
Manufacturing Operations
 Management (MOM), 99
Markram, Henry, 258
Mashey, John, 254
Mazor, Stanly, 62
McDermott, Terry, 281
McDevitt, Jack, 282
McMillan, Greg, 227
Measurement module, 189
Media access control (MAC), 140
Merton, Robert, 283
Mesh networks, 163

Minorsky, Nicolas, 49
Minsky, Marvin, 279
MIS. *See* Management Information
 Systems
Mitereff, S. D., 49
Model Predictive Control (MPC), 67
Modicon 184, 170
MOM. *See* Manufacturing Operations
 Management
Moravec, Hans, 278
Morley, Dick, 59
MPC. *See* Model Predictive Control
Mumford, Lewis, 3

Nanomaterials, 249
NE 124, 165
Neitzel, Lee, 65
Neumann, John von, 51
Neural networks, 65
Neurons, 279
Newcomen, Thomas, 3
Nexialist, 228
Nimmo, Ian, 264
Nodes, 139
Norman, D. A., 119
Nuisance alarm, 84

Object Dictionary (OD), 155
Octet, 139
OD. *See* Object Dictionary
ODVA. *See* Open DeviceNet Vendors
 Association
OLE for Process Control
 (OPC), 175
Oliver, Bernard, 285
Opaque automation, 124
OPC. *See* OLE for Process Control
OPC Unified Architecture
 (OPC UA), 176
OPC UA. *See* OPC Unified
 Architecture
Open DeviceNet Vendors Association
 (ODVA), 173
Open Platform Communications, 176

THIS TITLE IS FROM OUR MANUFACTURING COLLECTION. OTHER TITLES OF INTEREST MIGHT BE...

Announcing Digital Content Crafted by Librarians

Momentum Press offers digital content as authoritative treatments of advanced engineering topics, by leaders in their fields. Hosted on ebrary, MP provides practitioners, researchers, faculty and students in engineering, science and industry with innovative electronic content in sensors and controls engineering, advanced energy engineering, manufacturing, and materials science. **Momentum Press offers library-friendly terms:**

- perpetual access for a one-time fee
- no subscriptions or access fees required
- unlimited concurrent usage permitted
- downloadable PDFs provided
- free MARC records included
- free trials

The **Momentum Press** digital library is very affordable, with no obligation to buy in future years.

For more information, please visit **www.momentumpress.net/library** or to set up a trial in the US, please contact **mpsales@globalepress.com.**

www.ingramcontent.com/pod-product-compliance
Lightning Source LLC
Chambersburg PA
CBHW082003190326
41458CB00010B/3057